Phytoremediation *of* Hydrocarbon-Contaminated Soil

Phytoremediation *of* Hydrocarbon-Contaminated Soil

Edited by

Stephanie Fiorenza
Rice University, Houston, TX

Carroll L. Oubre
Rice University, Houston, TX

C. Herb Ward
Rice University, Houston, TX

Authors
M.K. Banks
R.S. Govindaraju
A.P. Schwab
P. Kulakow
J. Finn

CRC Press
Taylor & Francis Group
Boca Raton London New York

CRC Press is an imprint of the
Taylor & Francis Group, an **informa** business

CRC Press
Taylor & Francis Group
6000 Broken Sound Parkway NW, Suite 300
Boca Raton, FL 33487-2742

First issued in paperback 2019

© 2000 by Taylor & Francis Group, LLC
CRC Press is an imprint of Taylor & Francis Group, an Informa business

No claim to original U.S. Government works

ISBN-13: 978-1-56670-463-2 (hbk)
ISBN-13: 978-0-367-39954-2 (pbk)

Visit the Taylor & Francis Web site at
http://www.taylorandfrancis.com

and the CRC Press Web site at
http://www.crcpress.com

Foreword

Interest in phytoremediation has exploded since the principal investigators proposed this technology demonstration to the Advanced Applied Technology Demonstration Facility (AATDF) sponsored by the Department of Defense. Phytoremediation is being studied as a solution for a variety of contaminants in groundwater and soil, using different types of vegetation. The investigators from Kansas State University (now with Purdue University) focused their attention on the problem of hydrocarbon contamination, especially polynuclear aromatic hydrocarbons, in surface and near-surface soils.

The AATDF was interested in this particular project because of its attention to statistical design and analysis. One of the goals of the AATDF was to produce quantified results. With the randomized block design used in the phytoremediation field demonstration, the investigators were able to test whether there was a statistically significant difference in the effects of different vegetated and unvegetated treatments.

The field demonstration was conducted at the U.S. Navy's Craney Island Fuel Terminal in Portsmouth, VA, in soils contaminated with aged diesel fuel. The degradation of diesel and polynuclear aromatic hydrocarbon components was tested in plots containing three different vegetation treatments, two grasses and a legume, and a nonvegetated control. This field demonstration would not have been possible without the acceptance and aid of personnel at the U.S. Navy, Atlantic Division, who manage the environmental work at Craney Island Fuel Terminal. Special thanks go to Mr. Ernie Lory of the Naval Facilities Engineering Service Center, who first identified the field site for AATDF.

Part I of this monograph relates the results of the field demonstration. The reader will find it to be a complete and thorough account. Part II of the monograph covers the design and the potential costs of a full-scale implementation of the demonstration system at the Craney Island site, as well as operating parameters and other applications for this technology. This section was prepared by Remediation Technologies, Inc.

The authors would like to acknowledge Susan Pekarek, Hadasse Mallede, Jinbo Su, Charles Gray, Heather Leason, Jennifer Vosler, Steven Lillehaug, Jeff Keller, Kenchong Tan, and Andy Buesing, all from Kansas State University, for their assistance with laboratory analyses. The authors worked closely with the AATDF staff and project advisors throughout the demonstration. The project advisors were Mr. Richard Conway, Union Carbide Corporation; Mr. Joseph Kreitinger, Remediation Technologies, Inc.; Mr. Eric Chai and Dr. Ileana Rhodes, Shell Westhollow Technology Center; and Dr. Scott Cunningham, DuPont. The Technology Advisory Board and other AATDF advisory committees provided valuable assistance in project selection, progress review, and technology transfer. Dr. Karen Duston and Mr. Richard Conway served as the final reviewers and facilitated the publication process.

The AATDF Program was funded by the U.S. Department of Defense under Grant No. DACA39-93-1-001 to Rice University. The program was given oversight by the U.S. Army Engineer Waterways Experiment Station in Vicksburg, MS.

AATDF Monographs

This monograph is one of a ten-volume series that record the results of the AATDF Program:

- Surfactants and Cosolvents for NAPL Remediation (A Technology Practices Manual)
- Sequenced Reactive Barriers for Groundwater Remediation
- Modular Remediation Testing System
- Phytoremediation of Hydrocarbon-Contaminated Soil
- Steam Remediation of Contaminated Soils
- Soil Vapor Extraction: Radio Frequency Heating (Resource Manual and Technology Demonstration)
- Subsurface Contamination Monitoring Using Laser Fluoresence
- Reuse of Surfactants and Cosolvents for NAPL Remediation
- Remediation of Firing-Range Impact Berms
- Surfactant/Cosolvent Enhanced Subsurface Remediation

Advanced Applied Technology Demonstration Facility (AATDF)
Energy and Environmental Systems Institute MS-316
Rice University
6100 Main Street
Houston, TX 77005-1892

U.S. Army Engineer
Waterways Experiment Station
3909 Halls Ferry Road
Vicksburg, MS 39180-6199

RICE

Rice University
6100 Main Street
Houston, TX 77005-1892

Preface

Following a national competition, the Department of Defense (DOD) awarded a $19.3 million grant to a university consortium of environmental research centers led by Rice University and directed by Dr. C. Herb Ward, Foyt Family Chair of Engineering. The DOD Advanced Applied Technology Demonstration Facility (AATDF) Program for Environmental Remediation Technologies was established on May 1, 1993, to enhance the development of innovative remediation technologies for DOD by facilitating the process from academic research to full-scale utilization. The AATDF's focus is to select, test, and document performance of innovative environmental technologies for the remediation of DOD sites.

Participating universities include: Stanford University, The University of Texas at Austin, Rice University, Lamar University, University of Waterloo, and Louisiana State University. The directors of the environmental research centers at these universities serve as the Technology Advisory Board. The U.S. Army Engineer Waterways Experiment Station manages the AATDF grant for DOD. Dr. John Keeley is the Technical Grant Officer. The DOD/AATDF is supported by five leading consulting engineering firms: Remediation Technologies, Inc., Battelle Memorial Institute, GeoTrans, Inc., Arcadis Geraghty and Miller, Inc., and Groundwater Services, Inc., along with advisory groups from the DOD, industry, and commercialization interests.

Starting with 170 preproposals that were submitted in response to a broadly disseminated announcement, 12 projects were chosen by a peer review process for field demonstrations. The technologies chosen were targeted at DOD's most serious problems of soil and groundwater contamination. The primary objective was to provide more cost-effective solutions, preferably using *in situ* treatment. Eight projects were led by university researchers, two projects were managed by government agencies, and two others were conducted by engineering companies. Engineering partners were paired with the academic teams to provide field demonstration experience. Technology experts helped guide each project.

DOD sites were evaluated for their potential to support quantitative technology demonstrations. More than 75 sites were evaluated in order to match test sites to technologies. Following the development of detailed work plans, carefully monitored field tests were conducted, and the performance and economics of each technology were evaluated.

One AATDF project designed and developed two portable Experimental Controlled Release Systems (ECRS) for testing and field simulations of emerging remediation concepts and technologies. The ECRS is modular and portable and allows researchers, at their sites, to safely simulate contaminant spills and study remediation techniques without contaminant loss to the environment. The completely contained system allows for accurate material and energy balances.

The results of the DOD/AATDF Program provide the DOD and others with detailed performance and cost data for a number of emerging, field-tested technologies. The program also provides information on the niches and limitations of the technologies to allow for more informed selection of remedial solutions for environmental cleanup.

The AATDF Program can be contacted at Energy and Environmental Systems Institute, MS-316, Rice University, 6100 Main, Houston, TX 77005, phone 713-527-4700, fax 713-285-5948, e-mail <eesi@rice.edu>.

The DOD/AATDF Program staff includes:

Director:
  Dr. C. Herb Ward
Program Manager:
  Dr. Carroll L. Oubre
Assistant Program Manager:
  Dr. Kathy Balshaw-Biddle
Assistant Program Manager:
  Dr. Stephanie Fiorenza

Assistant Program Manager:
  Dr. Donald F. Lowe
Financial/Data Manager:
  Mr. Robert M. Dawson
Publications Coordinator/Graphic Designer:
  Ms. Mary Cormier
Meeting Coordinator:
  Ms. Susie Spicer

This volume, *Phytoremediation of Hydrocarbon-Contaminated Soil*, is part of a ten-monograph series that records the results of the DOD/AATDF environmental technology demonstrations. Many have contributed to the success of the AATDF program and to the knowledge gained. We trust that our efforts to fully disclose and record our findings will contribute valuable lessons learned and help further innovative technology development for environmental cleanup.

Stephanie Fiorenza
Carroll L. Oubre
C. Herb Ward

Editors

Stephanie Fiorenza

Stephanie Fiorenza is an Assistant Program Manager with AATDF at Rice University, where she has managed five projects involving the field demonstration of innovative remediation technologies. Dr. Fiorenza has a Ph.D. in environmental science and engineering from Rice University and a B.A. in environmental studies from Brown University. In her role as an assistant program manager for AATDF, Dr. Fiorenza provided the managerial guidance and technical expertise that were required for the successful field demonstration of each technology. She has also been an active participant in the preparation of all the project reports. Prior to joining AATDF, Dr. Fiorenza worked as an environmental specialist for Amoco Corporation in its Groundwater Management Section. Her areas of interest are the biodegradation of organic chemicals, microbial ecology of the subsurface, and development of remediation technologies.

Carroll L. Oubre

Carroll L. Oubre is the Program Manager for the DOD/AATDF Program. As Program Manager, he is responsible for the day-to-day management of the $19.3 million DOD/AATDF Program. This includes guidance of the AATDF staff, overview of the 12 demonstration projects, and assuring that project milestones are met within budget and that complete reporting of the results is timely.

Dr. Oubre has a B.S. in chemical engineering from the University of Southwestern Louisiana, an M.S. in chemical engineering from Ohio State University, and a Ph.D. in chemical engineering from Rice University. He worked for Shell Oil Company for 28 years, where his last job was Manager of Environmental Research and Development for Royal Dutch Shell in England. Prior to that, he was Director of Environmental Research and Development at Shell Development Company in Houston, TX.

C.H. (Herb) Ward

C.H. (Herb) Ward is the Foyt Family Chair of Engineering in the George R. Brown School of Engineering at Rice University. He is also Professor of Environmental Science and Engineering and Ecology and Evolutionary Biology.

Dr. Ward has undergraduate (B.S.) and graduate (M.S. and Ph.D.) degrees from New Mexico State University and Cornell University, respectively. He also earned the M.P.H. in environmental health from the University of Texas.

Following 22 years as Chair of the Department of Environmental Science and Engineering at Rice University, Dr. Ward is now Director of the Energy and Environmental Systems Institute, a university-wide program designed to mobilize industry, government, and academia to focus on problems related to energy production and environmental protection.

Dr. Ward is also Director of the DOD/AATDF Program, a distinguished consortium of university-based environmental research centers supported by consulting environmental engineering firms in guiding the selection, development, demonstration, and commercialization of advanced applied environmental restoration technologies for the Department of Defense. For the past 18 years, he has directed the activities of the National Center for Ground Water Research, a consortium of universities charged with conducting long-range exploratory research to help anticipate and solve the nation's emerging groundwater problems. He is also Co-Director of the EPA-sponsored Hazardous Substances Research Center/South and Southwest, which focuses its research on contaminated sediments and dredged materials.

Dr. Ward has served as president of both the American Institute of Biological Sciences and the Society for Industrial Microbiology. He is the founding and current editor-in-chief of the international journal *Environmental Toxicology and Chemistry*.

Authors

M. Kathy Banks

M. Kathy Banks is currently an Associate Professor of Civil Engineering at Purdue University and was previously a faculty member at Kansas State University. After receiving her B.S. in environmental engineering from the University of Florida, she completed an M.S. at the University of North Carolina and a Ph.D. at Duke University. Her research focuses on phytoremediation and microbial characterization of soils contaminated with hazardous organic compounds. Dr. Banks is currently an associate editor of the *Journal of Phytoremediation* and has published over 30 articles and book chapters.

R.S. Govindaraju

R.S. Govindaraju is currently an Associate Professor in the Civil Engineering Department at Purdue University and was previously affiliated with Kansas State University. His research focus is in the areas of surface and subsurface modeling of water and contaminant transport. His early work was on the problem of mathematical modeling of overland flows. This work was followed by field-scale studies of solute transport in porous media, in which he studied the influence of spatial averaging of the flow and transport equations. His recent research interest is in problems dealing with spatial variability and scaling, both of which have hampered conventional methods of analysis. Some of his current research work attempts to understand and quantify spatial variability through such tools as geostatistics.

A. Paul Schwab

A. Paul Schwab is Associate Professor of Soil Physical Chemistry at Purdue University. After receiving a B.S. in chemistry and mineral engineering from the Colorado School of Mines, he attended Colorado State University and earned M.S. and Ph.D. degrees in soil chemistry. Dr. Schwab was employed for 2 years as a Research Scientist at Battelle Pacific Northwest Laboratories, followed by 15 years as a member of the faculty at Kansas State University. He moved to Purdue University in January 1998, where he is focusing his research on phytoremediation of petroleum-contaminated soils and the chemistry of heavy metals in soil and water. Dr. Schwab has served as associate editor of the *Soil Science Society of America Journal* and currently is associate editor of the *Journal of Environmental Quality*.

Peter A. Kulakow

Peter A. Kulakow is a Research Associate in the Department of Agronomy at Kansas State University, where he provides plant science and horticultural support for phytoremediation projects. He has Ph.D. and B.S. degrees in genetics from the University of California at Davis. Dr. Kulakow has conducted applied genetic research in a number of areas, including new crop development, sustainable agriculture, applied ecology, and phytoremediation. He has a specific interest in selection of plant materials for solving environmental problems. Dr. Kulakow has developed a screening procedure for assessment of plant adaptation to petroleum-hydrocarbon-contaminated soils and has used this method to develop suitable mixtures of plant genotypes for revegetation and phytoremediation of hazardous waste sites.

John Finn

John Finn is an engineer and project manager at Remediation Technologies, Inc. with over 14 years of experience in environmental engineering. His areas of expertise are remedial design and construction at manufactured gas plant sites, sediment/soil remediation, bioremediation/phytoremediation, and solidification/stabilization. His B.S. is in chemical engineering from the University of Connecticut and his master's degree is in agricultural engineering from Cornell University.

John Finn

John Finn is an engineer and project manager at Remediation Technologies, Inc. with over 14 years of experience in environmental engineering. His areas of expertise are remedial design and construction at manufactured gas plant sites, sediment/soil remediation, bioremediation/phytoremediation, and solidification/stabilization. His B.S. is in chemical engineering from the University of Connecticut and his master's degree is in agricultural engineering from Cornell University.

Authors

M. Kathy Banks

M. Kathy Banks is currently an Associate Professor of Civil Engineering at Purdue University and was previously a faculty member at Kansas State University. After receiving her B.S. in environmental engineering from the University of Florida, she completed an M.S. at the University of North Carolina and a Ph.D. at Duke University. Her research focuses on phytoremediation and microbial characterization of soils contaminated with hazardous organic compounds. Dr. Banks is currently an associate editor of the *Journal of Phytoremediation* and has published over 30 articles and book chapters.

R.S. Govindaraju

R.S. Govindaraju is currently an Associate Professor in the Civil Engineering Department at Purdue University and was previously affiliated with Kansas State University. His research focus is in the areas of surface and subsurface modeling of water and contaminant transport. His early work was on the problem of mathematical modeling of overland flows. This work was followed by field-scale studies of solute transport in porous media, in which he studied the influence of spatial averaging of the flow and transport equations. His recent research interest is in problems dealing with spatial variability and scaling, both of which have hampered conventional methods of analysis. Some of his current research work attempts to understand and quantify spatial variability through such tools as geostatistics.

A. Paul Schwab

A. Paul Schwab is Associate Professor of Soil Physical Chemistry at Purdue University. After receiving a B.S. in chemistry and mineral engineering from the Colorado School of Mines, he attended Colorado State University and earned M.S. and Ph.D. degrees in soil chemistry. Dr. Schwab was employed for 2 years as a Research Scientist at Battelle Pacific Northwest Laboratories, followed by 15 years as a member of the faculty at Kansas State University. He moved to Purdue University in January 1998, where he is focusing his research on phytoremediation of petroleum-contaminated soils and the chemistry of heavy metals in soil and water. Dr. Schwab has served as associate editor of the *Soil Science Society of America Journal* and currently is associate editor of the *Journal of Environmental Quality*.

Peter A. Kulakow

Peter A. Kulakow is a Research Associate in the Department of Agronomy at Kansas State University, where he provides plant science and horticultural support for phytoremediation projects. He has Ph.D. and B.S. degrees in genetics from the University of California at Davis. Dr. Kulakow has conducted applied genetic research in a number of areas, including new crop development, sustainable agriculture, applied ecology, and phytoremediation. He has a specific interest in selection of plant materials for solving environmental problems. Dr. Kulakow has developed a screening procedure for assessment of plant adaptation to petroleum-hydrocarbon-contaminated soils and has used this method to develop suitable mixtures of plant genotypes for revegetation and phytoremediation of hazardous waste sites.

AATDF Advisors

University Environmental Research Centers

National Center for Ground Water Research
Dr. C. H. Ward
Rice University, Houston, TX

Hazardous Substances Research Center–South and Southwest
Dr. Danny Reible and
Dr. Louis Thibodeaux
Louisiana State University
Baton Rouge, LA

Waterloo Centre for Groundwater Research
Dr. John Cherry and Mr. David Smyth
University of Waterloo, Ontario, Canada

Western Region Hazardous Substances Research Center
Dr. Perry McCarty
Stanford University, Stanford, CA

Gulf Coast Hazardous Substances Research Center
Dr. Jack Hopper and Dr. Alan Ford
Lamar University, Beaumont, TX

Environmental Solutions Program
Dr. Raymond C. Loehr
University of Texas, Austin, TX

DOD/Advisory Committee

Dr. John Keeley, Co-Chair
Assistant Director,
Environmental Laboratory
U.S. Army Corps of Engineers
Waterways Experiment Station
Vicksburg, MS

Mr. James I. Arnold, Co-Chair
Acting Division Chief, Technical Support
U.S. Army Environmental Center
Aberdeen, MD

DOD/Advisory Committee (continued)

Dr. John M. Cullinane
Program Manager, Installation Restoration
Waterways Experiment Station
U.S. Army Corps of Engineers
Vicksburg, MS

Mr. Scott Markert and Dr. Shun Ling
Naval Facilities Engineering Center
Alexandria, VA

Dr. Jimmy Cornette, Dr. Michael Katona and Major Mark Smith
Environics Directorate
Armstrong Laboratory
Tyndall AFB, FL

Commercialization and Technology Transfer Advisory Committee

Mr. Benjamin Bailar, Chair
Dean, Jones Graduate School
of Administration
Rice University, Houston, TX

Dr. James H. Johnson, Jr., Associate Chair
Dean of Engineering
Howard University, Washington, D.C.

Dr. Corale L. Brierley
Consultant
VistaTech Partnership, Ltd.
Salt Lake City, UT

Dr. Walter Kovalick
Director, Technology Innovation Office
Office of Solid Wastes and Emergency
Response
U.S. EPA, Washington, D.C.

Mr. M.R. (Dick) Scalf (retired)
U.S. EPA
Robert S. Kerr Environmental
Research Laboratory
Ada, OK

Contents

Part I: Field Demonstration

M.K. Banks, R.S. Govindaraju, A.P. Schwab, P. Kulakow

Contents

Part I: Field Demonstration

M.K. Banks, R.S. Govindaraju, A.P. Schwab, P. Kulakow

Part II: Technology Design/Evaluation

J. Finn

List of Tables

List of Figures

Acronyms and Abbreviations

AATDF	Advanced Applied Technology Development Facility
ANOVA	analysis of variance
ASA	American Society of Agronomy
ASTM	American Society for Testing and Materials
BA	benzo[a]anthracene
BaP	benzo[a]pyrene
BTEX	benzene, toluene, ethylbenzene, xylenes
C	carbon
^{14}C	radioactive isotope of carbon with mass number 14
°C	degrees centigrade
CEC	cation exchange capacity
CFU	colony-forming unit
CIFT	Craney Island Fuel Terminal
cm	centimeter
CO_2	carbon dioxide
DOD	Department of Defense
dS/m	unit of electrical conductivity
EC	electrical conductivity
EPA	Environmental Protection Agency
FID	flame ionization detection/detector
FRTR	Federal Remediation Technologies Roundtable
ft	feet
g	gram
gpm	gallons per minute
GC	gas chromatography
HCl	hydrochloric acid
ID	internal diameter
IR	infrared (spectrometry)
$K_2Cr_2O_7$	potassium dichromate
kg	kilogram
KSU	Kansas State University
LSD	least significant difference
m	meter
mg	milligram
min	minute
ml	milliliter
mm	millimeter
m/m	mass/mass
MPN	most probable number
MS	mass spectrometry
MSD	mass selective detector
m/z	mass/charge ratio
N	nitrogen
NaOH	sodium hydroxide
Na_2SO_4	sodium sulfate
$Na_4P_2O_7$	tetrasodium pyrophosphate
nd	not detected

NH_4^+	ammonium ion
$(NH_4)_2HPO_4$	ammonium phosphate
nm	nanometer
NO_3^-	nitrate ion
P	phosphorus
p	probability
PAH	polynuclear aromatic hydrocarbon
PCP	pentachlorophenol
ppm	parts per million
PVC	polyvinyl chloride
RBCA	risk-based cleanup action
rpm	revolutions per minute
SAR	sodium adsorption ratio
SD	standard deviation
SFE	supercritical fluid extraction
SIM	selected ion monitoring
TER	Technical Evaluation Report
TOC	total organic carbon
TPH	total petroleum hydrocarbon
TSA	tryptic soy agar
μCi	microcurie
μg	microgram
μl	microliter
USDA	U.S. Department of Agriculture
v:v	volume:volume
WBS	work breakdown structure

PART I

Field Demonstration

M.K. Banks, R.S. Govindaraju, A.P. Schwab
Purdue University

P. Kulakow
Kansas State University

Executive Summary of Demonstration

Petroleum-derived hydrocarbon contamination of soil is a serious problem throughout the U.S. Bioremediation of petroleum in soil using indigenous microorganisms has proven effective; however, the biodegradation rate of more recalcitrant and potentially toxic petroleum contaminants, such as polycyclic aromatic hydrocarbons (PAHs), is rapid at first but declines quickly. Biodegradation of such compounds is limited by their strong adsorption potential and low solubility. Vegetation may play an important role in the biodegradation of toxic organic chemicals in soil. For petroleum compounds, the presence of rhizosphere microflora may accelerate biodegradation of the contaminants. The establishment of vegetation on hazardous waste sites is an economic, potentially effective, low-maintenance approach to waste remediation and stabilization.

In greenhouse studies, phytoremediation was found to be a feasible method for cleanup of surface soil contaminated with petroleum products. However, a field trial was needed to demonstrate this new technology. In this project, aged petroleum-contaminated soil located at the Craney Island Fuel Terminal Biological Treatment Facility was treated using phytoremediation. The Craney Island Fuel Terminal, the Navy's largest fuel facility in the U.S., was the location for this phytoremediation field study. This study was conducted on 0.5 acre (0.203 ha) of the Atlantic Division Naval Facilities Engineering Command biological treatment cell. Petroleum-contaminated soil was placed in the cell to a depth of approximately 2 ft (0.61 m). The soil was sampled and characterized, and plants were established in the phytoremediation area. The study area was divided into distinct plots of approximately the same size. There were six replicate plots per treatment. The vegetated plots contained one of the following plant types: rapidly growing cool-season grass with an intensive root system (fescue), a mixture of warm-season perennial grass species (Bermuda grass) and cool-season annual grass (rye grass), and a shallow-rooted legume (white clover). A series of analyses was performed over time to evaluate the efficiency of remediation and overall soil health.

All three vegetation treatments grew well in the contaminated soil, but the pattern of growth varied for each treatment (Table ES.1). Bermuda grass was established by sod and provided the most rapid cover and rooting for all species. Seeding perennial rye grass into the Bermuda grass in the fall helped provide winter growth and rooting while the Bermuda grass was dormant. By the 7-month vegetation sampling in May 1996, Bermuda grass/rye grass had the most root production of all treatments. Tall fescue also grew well throughout the study. Since established by seed, it took longer to establish a full root system. At the 12-month sampling in September 1996, tall fescue had the highest mass and density of roots for all treatments. Tall fescue also had a deeper rooting system than the other treatments. Both the tall fescue and Bermuda grass persisted well through the dry second growing season. Both treatments showed reduced root growth in the second year but maintained a complete canopy.

Microbial numbers and diversity were initially higher in the vegetated plots, but were approximately the same as the unvegetated plots in the second year of the study (Table ES.2). The clover plots, which had the highest number of microbial petroleum degraders and the

Table ES.1   Average Root (0- to 10-cm Depth) and Shoot Biomass in Phytoremediation Area

Treatment	g/m²			
	13 months		24 months	
	Root	Shoot	Root	Shoot
Bermuda	92	557	63	333
Fescue	104	298	54	463
Clover	111	195	10	—

lowest total petroleum hydrocarbon (TPH) concentrations at the last sampling event, also had the most root turnover. Sustained microbial degradation efficiency in the vegetated plots may be directly related to root turnover.

There was a statistically significant reduction of TPH in all vegetation plots as compared to the unvegetated controls (Table ES.3). White clover showed the highest rate of TPH degradation followed by tall fescue and Bermuda grass. Vegetated treatments had 9 to 19% increased degradation when compared to the unvegetated control plots. There was no evidence of a plateau having been reached at the 24-month sampling date (i.e., further reduction in TPH was anticipated). During the remediation process, the hydrocarbons did not leach from the root zone, and the plants did not accumulate PAHs in the shoots.

The trends observed in the TPH concentrations could not be directly transferred to PAH concentrations. For nearly all the individual PAHs studied, the percent degradation was greatest in the fescue plots and least in the unvegetated plots. Very often, the percent degradation of PAHs in the clover plots was statistically less than in the fescue and equivalent to that in the unvegetated plots. TPH degradation in the clover plots increased sharply when the clover was dying and the roots were degrading. The hydrocarbons may have been degraded cometabolically, impacting the more labile compounds first and the recalcitrant PAHs last. Fescue's fine root structure, in contrast with the coarse root structure of clover, can penetrate microsites in soil and promote degradation of a greater percentage of the PAHs.

The relationship between plant growth and hydrocarbon degradation in this study did not follow the simple pattern that greater plant growth resulted in greater degradation rates. Among the plant treatments, white clover had the lowest root and aboveground biomass production, yet it had the highest rate of hydrocarbon degradation. The white clover in this study grew well in the first 12 months of this study. However, due primarily to dry conditions in the second year, white clover did not thrive. A significant amount of the clover roots had decomposed by the final sampling event. Although the tall fescue and Bermuda grass continued to grow in the second year, these treatments also experienced significant root decomposition. In the first year of the study, it is possible that root growth and the influence of plant roots on soil physical properties and microbial activity were important to hydrocarbon

Table ES.2   Microbial Numbers, Petroleum Degraders, and BIOLOG Diversity in Phytoremediation Plots after 24 Months of Plant Growth

Treatment	Microbial Numbers (log CFU/g dry soil)[a]	Petroleum Degraders (MPN)[b]	BIOLOG Diversity (%)
Clover	6.73	3.85×10^7	30
Fescue	5.77	1.16×10^6	20
Bermuda	5.91	1.25×10^6	27
Bare	6.83	2.99×10^5	31

[a] CFU = colony-forming unit.

[b] MPN = most probable number.

Table ES.3 TPH Dissipation for the Craney Island Phytoremediation Field Study

Treatment	13 months	24 months
Clover	29%	50%
Fescue	33%	45%
Bermuda	27%	40%
Bare	21%	31%

degradation, while in the second year, rates of root decomposition may have played an increasingly important role. While it is necessary to establish healthy vegetation on phytoremediation sites, other factors such as the rate of root turnover, root exudation patterns, and the influence of vegetation on soil physical properties appear to be just as important as the quantity of vegetation produced. Environmental factors such as available water and soil conditions should be carefully managed to enhance the remediation process.

Several laboratory and greenhouse studies were conducted in support of the field research. A shaking extraction method was developed for analysis of aged petroleum contaminants in soil. This new method is quicker and uses less solvent than the standard soxhlet approach. Soxhlet extraction of wet samples, even when partially dried by adding anhydrous sodium sulfate, gave varying results. The improved method for meaningful TPH analysis involved complete drying, sieving, and mixing (by grinding) of a 1-g sample followed by three sequences of shaking extraction with dichloromethane. A plant growth chamber study was conducted to evaluate the fate of benzo[a]pyrene (BaP) in vegetated systems. Results indicate that BaP is degraded in the rhizosphere; however, degradation by-products ultimately become incorporated into the soil matrix. A greenhouse study was conducted to evaluate the effect of depth on phytoremediation efficiency. After 21 weeks of plant growth, vegetated columns had lower residual concentrations of benzo[a]anthracene and BaP in deeper soil layers than the unvegetated columns. These results imply that phytoremediation may have a significant impact on contaminants in shallow subsoil.

Future studies are needed to evaluate root turnover and associated phytoremediation efficiency. The plant roots appear to provide an ideal environment for degradation of organic compounds. Roots improve the structure of the soil by allowing rapid movement of water and gases through the soil. Microorganisms associated with roots may be present in regions of the soil that might otherwise be inaccessible, particularly the interior of dense aggregates and similar microsites. The rhizosphere encourages high microbial populations and activities by improving transport of water and air, providing elevated concentrations of labile carbon through sloughing of cells and root exudation, and allowing rapid achievement of near-ideal moisture contents by encouraging water flow through the soil profile and removal of excess water after heavy rainfall. A very fine, fibrous root system does not appear to be critical for phytoremediation; white clover has a coarse structure, yet the plots with clover had some of the highest rates of degradation.

Plant mixtures of grasses and legumes may be an important combination for optimal microbial activity. Root turnover and new root growth could impact the microbial responses for sustained TPH degradation. While it is important to understand the mechanisms responsible for vegetation-enhanced remediation of petroleum-hydrocarbon-contaminated soil, it is just as important to continue field-scale demonstrations of the technology to establish management protocols and supporting data to facilitate adoption of phytoremediation technology.

This technology has broad applicability when the conditions are conducive to adequate plant growth and if the contaminants of interest are accessible to the roots. Growing seasons of reasonable length, a nonphytotoxic contaminant matrix, a hospitable climate, and availability of a source of seeds or plants are some important considerations. Although the implementation of phytoremediation is not technically complex, a certain level of expertise is needed to ensure that the entire system is designed to provide the best opportunity for success. Whether the technology is being used as a final polishing step or as the primary remediation approach, the minimal amount of site management and low cost will make it an attractive alternative.

CHAPTER 1

Demonstration Introduction and Technology Overview

1.1 OVERALL PROJECT DESCRIPTION

Petroleum contamination of soil is a serious problem throughout the U.S. Bioremediation of petroleum in soil using indigenous microorganisms has proven effective; however, the biodegradation rate of more recalcitrant and potentially toxic petroleum contaminants, such as polycyclic aromatic hydrocarbons (PAHs), is rapid at first but declines quickly. Biodegradation of such compounds is limited by their strong adsorption potential and low solubility. Vegetation may play an important role in the biodegradation of toxic organic chemicals in soil. For petroleum compounds, the presence of rhizosphere microflora may accelerate biodegradation of the contaminants. The establishment of vegetation on hazardous waste sites is an economic, potentially effective, low-maintenance approach to waste remediation and stabilization.

In greenhouse studies, phytoremediation was found to be a feasible method for cleanup of surface soil contaminated with petroleum products. However, a field trial was needed to demonstrate this new technology. In this project, aged petroleum-contaminated soil located at the Craney Island Fuel Terminal Biological Treatment Facility was treated using phytoremediation.

The demonstration field site was a small area (approximately 0.5 acre or 0.2 ha) located in the Craney Island Fuel Terminal biotreatment cell, where different vegetation regimes suitable to the local area were evaluated for their phytoremediation potential. The study area was divided into distinct plots of approximately the same size. The vegetated plots contained one of the following plant types: rapidly growing cool-season grass with an intensive root system (tall fescue [*Festuca arundinacea* var. Kentucky 31]), a mixture of warm-season perennial grass species (Bermuda grass [*Cynodon dactylon* var. Vamont]) and cool-season grass (perennial rye grass [*Lolium perene* var. Linn]), and a shallow-rooted legume (white clover [*Trifolium repens* var. Dutch White]). These plants were chosen based on the germination studies and consultation with local experts. Unvegetated plots were used as controls. Plots were seeded in amounts typical for each species or sodded. Fertilizer applications were performed by Kansas State University (KSU) personnel initially and periodically during the study. The rates of fertilization were chosen to provide a C:N:P ratio of 100:20:10 over the course of the study period. This amount provides the N and P needed for plant growth plus the nutrients necessary for soil microorganisms to degrade the contaminants. KSU and OHM Remediation Services Corporation personnel applied water using an irrigation system supplied by KSU with water available at the site.

The primary means of assessing contaminant dissipation was soil analysis using an improved extraction procedure. Throughout the growing season, soils were sampled. The samples were analyzed for total petroleum hydrocarbons (TPHs) and microbial characteristics at KSU laboratories. Split soil samples were sent to a contract laboratory for TPH analysis. Plant root and shoot biomass was assessed. Also, soil solution samplers (vacuum pore-water samplers) were installed in the phytoremediation plots to assess leaching of contaminants. These samplers collected soil solution directly above the underlying sand layer. This monitoring continued for 2 years. Detailed results from these analyses can be found in Section 3.5.

1.2 REVIEW OF PHYTOREMEDIATION

Recently published research has indicated that vegetation may play an important role in the bioremediation of toxic organic chemicals (Walton and Anderson, 1990; Lappin et al., 1985; Sandmann and Loos, 1984; Hsu and Bartha, 1979; Reddy and Sethunanthan, 1983; Anderson and Walton, 1991; Aprill and Sims, 1990; Ferro et al., 1994; Reilley et al., 1996). The establishment of vegetation on moderately contaminated hazardous waste sites may be an effective and low-maintenance approach to waste remediation. The use of plants for remediation may be especially well suited for soils contaminated by organic chemicals to depths of less than 2 m (Bell, 1992). The results of research cited above indicate that the interaction between plant roots and rhizosphere microflora significantly enhances degradation of hazardous organic compounds in contaminated soil.

Plants can interact with hazardous organic compounds through degradation or accumulation (Finlayson and MacCarthy, 1973). Uptake of the contaminant by the root is a direct function of the pollutant concentration in the soil solution and usually involves chemical partitioning on the root surfaces followed by movement across the cortex to the plant's vascular system (Crowdy and Jones, 1956). The contaminant may be bound or metabolized at any point during transport. Contaminants may be found in plants as freely extractable residues, extractable conjugate bound to plant material, and unextractable residues incorporated in plant tissue (Bell and Failey, 1991). Within a plant, the contaminant may be adsorbed on a cell surface or accumulated in the cell. Many contaminants become bound on the root surface and are not translocated (Bell, 1992).

Plants may indirectly contribute to the dissipation of contaminants in vegetated soil. Soil adjacent to the root contains increased microbial numbers and populations (Paul and Clark, 1989). Rovira and Davey (1974) found the number of bacteria in the rhizosphere to be as much as 20 times that normally found in nonrhizosphere soil. Short, Gram-negative rods (specifically *Pseudomonas*, *Flavobacterium*, and *Alcaligenes*) are most commonly found in the rhizosphere (Barber, 1984). The increased microbial numbers are primarily due to the presence of plant exudates and sloughed tissue which serve as sources of energy, carbon, nitrogen, or growth factors. The products excreted by plants include amino acids, carboxylic acids, carbohydrates, nucleic acid derivatives, growth factors, enzymes, and other related compounds (Alexander, 1977). The activity of microorganisms in the root zone stimulates root exudation, which further stimulates microbial activity (Barber and Martin, 1976).

Several reported studies have evaluated the effect of plants and the associated rhizosphere on the fate of petroleum contaminants (Reilley et al., 1996; Ferro et al., 1994; Schwab and Banks, 1994; Aprill and Sims, 1990). For the most part, the presence of plants enhanced the dissipation of the contaminants. Also, in the studies using ^{14}C-labeled contaminants in closed plant chambers, mineralization was greater in rhizosphere soils than in unvegetated soil, indicating that the bioavailability of the contaminant was increased in the rhizosphere.

Aprill and Sims (1990) studied the effects of using deep-rooted prairie grasses to remediate soil contaminated with PAHs. They suggested that the roots of these perennial grasses may be more effective at stimulating the rhizosphere microflora due to their fibrous nature. Fibrous roots offer more root surface area for microbial colonization than other roots and result in a larger microbial population in the contaminated soil. Big bluestem (*Andropogon gerardii*), Indian grass (*Sorghastrum nutans*), switchgrass (*Panicum virgatum*), Canada wild rye (*Elymus canadensis*), little bluestem (*Schizachyrium scoparius*), side oats grama (*Bouteloua curtipendula*), western wheatgrass (*Agropyron smithii*), and blue grama (*Bouteloua gracilis*) were evaluated. After 219 days of growth, PAH dissipation was greater in the rhizosphere soil when compared to nonrhizosphere soil. The dissipation ranking among the four PAHs studied correlated with the water solubility of the compound, with the more soluble compound showing the highest degradation. Ferro et al. (1994) evaluated the mineralization of pentachlorophenol (PCP) and phenanthrene in rhizosphere soil in a ^{14}C study. Crested wheatgrass (*Agropyron desertorum*) was grown in flow-through plant chambers. In the PCP-contaminated soil, 22% of the initial PCP was converted to $^{14}CO_2$ after 155 days in the vegetated soil, while only 5% was mineralized in the unvegetated soil.

Schwab and Banks (1994) investigated the degradation of PAHs in the rhizosphere of a variety of plants grown in petroleum-contaminated soil. Degradation in the contaminated soils was assessed for different plant types and PAH compounds. Alfalfa (*Meticago sativa*), fescue (*Festuca arundinacea*), big bluestem (*Andropogon gerardii*), and sudan grass (*Sorghum vulgare sudanense*) were used. Pyrene was one of the target compounds assessed in this study. After 4 weeks of plant growth, the concentration of pyrene had declined from an initial level of 100 mg/kg to <12.6 mg/kg. By 24 weeks, concentrations were <2.4 mg/kg in the soil. Pyrene degradation was significantly greater in the presence of plants than without plants. There were no detectable target PAHs (<10 mg/l) found in any of the leachates or plant biomass. Neither nonextractable pyrene nor nonbiological degradation could account for the observed changes in pyrene concentration. As indicated by these results, plant uptake, irreversible adsorption, abiotic degradation, and leaching could not account for any of the observed dissipation of pyrene. Therefore, biological degradation by microbes in the rhizosphere is the most likely explanation for the enhanced dissipation.

Based on evidence from greenhouse and field studies, phytoremediation is a viable remediation method for petroleum-contaminated soil. The use of vegetation for remediation of contaminated sites is attractive because it is inexpensive and passive. With few inputs and little management, a successful vegetation remediation system could be superior to many alternative cleanup techniques.

CHAPTER 2

Laboratory and Greenhouse Studies

Four laboratory and greenhouse studies were executed in support of this project. The first study was the development of a mechanical shaking method for the extraction of petroleum hydrocarbons from contaminated soil. The standard U.S. Environmental Protection Agency (EPA) method, soxhlet extraction, is labor intensive, requires expensive equipment, and was not developed for soils. A more rapid and cost-effective method for total petroleum hydro-carbon (TPH) extraction was needed in this project. The second study was a greenhouse investigation of degradation in the rhizosphere of benzo[*a*]pyrene, one of the most toxic compounds in petroleum-contaminated soils. The objective was to quantify the impact of plant roots on degradation and to attempt to find the ultimate fate of the degraded compound (e.g., complete mineralization to CO_2, conversion to volatile products, incorporation into organic matter). The third project was designed to quantify the degradation of diesel fuel and certain polycyclic aromatic hydrocarbons (PAHs) as a function of depth and time in the presence and absence of plants. Tall fescue was grown in tall greenhouse pots containing 60 cm of diesel-contaminated soil, and the soils were sampled at regular time intervals at three soil depths. The objective of this third study was to answer important questions about the effective depth of phytoremediation and the progression of the roots. The fourth study was a germination assessment of the Craney Island soil.

2.1 SHAKING METHOD FOR EXTRACTION OF PETROLEUM HYDROCARBONS FROM SOIL

Petroleum hydrocarbons are common soil contaminants from a variety of sources includ-ing leaking fuel storage tanks, crude oil spills, and production waste products. Chemical analysis of soil plays a crucial role in evaluating contaminated soils, and solvent extraction is a critical step. A rapid and reliable extraction method is needed to accurately analyze large numbers of soil samples.

Soxhlet extraction, an accepted protocol for extraction of semivolatile and nonvolatile organic compounds from soil matrices, has been outlined in detail in EPA Method 3540A (U.S. EPA, 1994). The disadvantages of soxhlet extraction are (1) the soil sample is static during the extraction process, which may limit contact between solvent and soil micropores; (2) soxhlet extraction requires up to 24 hr of extraction and specialized apparatus, which may be prohibitive for the analysis of large numbers of samples; and (3) high moisture content in soil samples may increase variability in the analysis because of the difficulty in obtaining representative subsamples and the poor interaction between nonpolar solvents and hydrated soil surfaces.

Several alternatives to soxhlet extraction of soils have been developed, and some have become accepted protocols. Sonication (EPA Method 3550/3550B) may be used interchangeably with soxhlet and has been tested for soils (Brilis and Marsden, 1990; Sawyer, 1996; Chen et al., 1996). This method consumes large quantities of solvent (similar to soxhlet), is labor intensive, and requires specialty equipment. Eckert-Tilotta and Hawthorne (1993) used a supercritical fluid extraction (SFE) method to extract TPHs in soil. This method is more rapid than soxhlet and eliminates the use of organic solvents. Unfortunately, SFE instrumentation is expensive and, when used to extract natural soil samples, may be subject to low accuracy and high variability (Reimer and Suarez, 1995). Accelerated solvent extraction, involving higher temperatures and pressures, was found to be generally equivalent to soxhlet extraction (Fisher et al., 1997).

A batch (shaking) extraction method (Chen et al., 1996) was compared to soxhlet and sonication. The method consisted of three sequential shaking extractions with 1:1 v:v methanol:dichloromethane; the first extraction was 3 days, and the final two were 18 hr. The extraction methods were found to be equivalent in the removal of PAHs and surrogate spikes. Shorter extraction times were not tested.

Organic solvents such as Freon, methanol, dichloromethane, and acetone can be used to extract semivolatile organic compounds from soil and other solid matrices. Freon use has been curtailed recently because of the potential environmental hazard. Methanol and acetone have intermediate polarity and are completely miscible with water. They can be used for extracting polar and nonpolar compounds from high-moisture-content matrices. Dichloromethane has relatively low polarity, is a good solvent for extracting nonpolar compounds when the sample has a low moisture content, and is well suited for gas-chromatographic analysis. However, dichloromethane and water may form an emulsion when the moisture content is high in the sample.

The overall objective of this study was to evaluate shaking methods for the extraction of volatile and semivolatile organic compounds from solids such as soils and soil-like materials. When several shaking cycles are employed, fresh solvent contacts the sample for each iteration of the sequence. Shaking ensures complete contact with the solid phase, thus increasing extraction efficiency. The potential advantages of this protocol are the use of simple and common equipment, reduction in the volume of organic solvents, extraction of many samples simultaneously, and cumulative extraction periods of less than 3 hr. Subobjectives were to (1) test several organic solvents with a range of properties to evaluate their ability to extract petroleum contaminants from soil; (2) determine if shaking with several consecutive aliquots of fresh solvent increases extraction efficiency; (3) examine the impact of aging and soil moisture on extraction of various petroleum contaminants for both field-aged soils and soils recently contaminated in the laboratory; and (4) determine the best shaking protocol for TPHs, PAHs, aliphatic hydrocarbons, branched alkanes, and other compounds of interest. Concentrations determined by shaking extraction were compared to those obtained by standard soxhlet extraction.

2.1.1 Materials and Methods

Four experiments were designed to test the efficacy of shaking extraction. In Experiment 1, three soils of differing texture were contaminated with PAHs, brought to a range of moisture contents, and extracted with several organic solvents. Experiment 2 was an extensive test of shaking vs. soxhlet on two soils that had been contaminated with petroleum hydrocarbons for at least 25 years. The effect of aging on extractability by shaking and soxhlet was tested in Experiment 3. Experiment 4 was the quantification of individual PAHs by gas chromatography/mass spectrometry after extraction by either shaking or soxhlet. Some of the experimental procedures were common to all three experiments, but others were specific to a given experiment, as described below.

2.1.1.1 Soil Samples and Study Sites

Uncontaminated soils that were amended with petroleum compounds for Experiments 1 and 3 were obtained in the vicinity of Manhattan, KS. For the test of solvents on soils with different textures, soils were chosen based on sand and clay content. Thus, the Sarpy soil (mixed, mesic, Typic Udipsamments) was chosen as a very sandy soil, the Eudora (coarse-silty, mixed mesic Fluventic Hapludolls) for intermediate texture, and the Smolan (fine, montmorillonitic mesic Pachic Argiustolls) for a high clay content (Table 2.1). Water contents at saturation were 16% for the Sarpy, 37% for Eudora, and 48% for the Smolan.

Contaminated soils (Experiment 2) were selected from two sites based on the type of contaminant, aging, and soil texture. One soil was from the Craney Island Fuel Terminal, Portsmouth, VA. This soil had been contaminated by diesel fuel for at least 50 years. Selected chemical and physical properties of this soil are given in Table 2.1. Although the electrical conductivity (EC) was elevated (4 dS/m) compared to normal soils (<1 dS/m) as a result of the contamination, none of the other properties were unusual. The second "soil" was an oxidation pond sludge from the Chevron Oil Refinery in Richmond, CA. It was a mixture of refinery sludge and sediment. As with the Virginia soil, the elevated EC (3.3 dS/m) was the only unusual soil test property (Table 2.1). The petroleum contaminants were a minimum of 25 years old and spanned the diesel and motor oil ranges of hydrocarbons.

Moist, contaminated soils from the two fields were homogenized and split into two subsamples. One part was dried at 30°C and ground to pass a 40-mesh sieve before extraction. The second part was passed through a 1-mm sieve, sealed in a glass bottle, and stored at 4°C to maintain moisture and reduce degradation.

2.1.1.2 Soil Contamination and Preparation

The purpose of Experiment 1 was to test the ability of several organic solvents to remove PAHs from soils of different textures. Uncontaminated, air-dried soil of each type was contaminated with anthracene, pyrene, and benzo[a]pyrene (BaP) to a concentration of approximately 30 mg/kg. A stock solution of the compounds was prepared by dissolving each in acetone to obtain a concentration of 500 mg/l. Each soil was contaminated by spraying 30 ml of the PAH stock solution in three equal parts to 500 mg of soil. The soil was mixed thoroughly between each addition to ensure uniform distribution of contaminants. The contaminated soils then were allowed to stand for 24 hr with intermittent stirring to allow complete evaporation of the acetone.

Table 2.1 Chemical and Physical Properties of the Soils Used in the Extraction Method Development

Property	Craney Island	Refinery Sludge	Sarpy	Eudora	Smolan
pH	7.4	7.5	8.1	8.0	6.7
Electrical conductivity (dS/m)	3.4	3.8	0.9	0.6	0.8
Cation exchange capacity (mmol/kg)	31	25	5.5	12	21
Organic C (g/kg)	11	10	0.1	0.8	2.4
Sand (%)	60	48	71	33	27
Silt (%)	21	28	26	57	46
Clay (%)	19	24	3	10	27

The moisture content of each soil was adjusted to 25, 50, and 100% of saturation by adding the appropriate amount of distilled water to each soil. The treatment for "air-dried" soils was obtained by wetting 500 g of each of the soils to 50% saturation and drying under ambient conditions in a greenhouse for 72 hr. Air-dry moisture contents were 0.9% (m/m) for Sarpy, 1.6% for Eudora, and 3.1% for Smolan.

The purpose of Experiment 3 was to determine the effects of aging on the extraction of specific organic compounds from soil by standard soxhlet or by shaking with acetone or dichloromethane. Eudora soil (Table 2.1) was selected for this study because of its interme- diate texture. The soil was contaminated with a synthetic diesel fuel containing 17 com- pounds (S. Cunningham, DuPont, personal communication), thoroughly homogenized, brought to 50% saturation, and allowed to age. A portion of the contaminated soil was weathered outdoors for 180 days in large containers to allow exposure to sunlight and rainfall. Another portion was aged only for 48 hr at 28°C. After their respective aging periods, both portions of contaminated soil were stored at 4°C for 180 days. Uncontaminated, moist soil was spiked with the synthetic diesel fuel immediately prior to extraction.

Soil samples from each of the three aging periods were extracted in triplicate by dichloromethane and soxhlet. The soils that had been aged and stored at 4°C were extracted moist (approximately 11% water by weight) or were ground with 25 g anhydrous sodium sulfate until the mixture flowed easily with no signs of clumping of moist soil. Uncontami- nated soil was ground with 25 g Na_2SO_4, transferred to a soxhlet thimble, spiked with the contaminant mixture, and the extraction began immediately.

2.1.1.3 Soil Extraction

In Experiment 1, five solvents were used to extract PAHs from three soil types. The solvents were chosen to represent a range of properties: hexane, dichloromethane, acetone, methanol, and methanol with 10% (v:v) water. The solvents used in Experiment 2 were dichloromethane, acetone, and 50:50/v:v dichloromethane and acetone. For Experiment 3, dichloromethane and acetone were used. All organic solvents were Optima Grade (Fisher Scientific, St. Louis, MO).

For soxhlet extraction, 5 g of each soil was extracted using 100 ml dichloromethane for 24 hr. Replication ranged from $n = 3$ to $n = 5$. The extraction solutions were sealed in glass vials and stored at 4°C until analysis.

For all shaking extractions, soils were weighed into 20-ml glass scintillation vials. The soil weight was 2.5 g in Experiments 1 and 3. Experiment 2 tested the effect of soil weight on extraction efficiency using 1, 3, or 5 g of soil. In all cases, 10 ml of solvent was added to each vial. The vials were sealed with a foil-lined cap and shaken for one extraction period with a mechanical reciprocal shaker at 160 cycles/min. The shaking period was 60 min for Experiment 1, 30 min for Experiment 3, and varied from 30 min to 4 hr for Experiment 2. After shaking, the scintillation vials were centrifuged for 10 min at 180 g. Scintillation vials were removed carefully from the centrifuge, and the extraction solution was decanted and saved.

When more than one extraction cycle was used, 10 ml of clean solvent was added to the soil used previously to begin a new extraction cycle. The process was repeated until the required number of cycles was completed. The extracts from all cycles for a given sample were combined and weighed. Thus, a sample extracted with only one cycle would have approximately 10 ml extract, a sample with two cycles would have approximately 20 ml, and so on. Extracts were stored at 4°C until analysis.

2.1.1.4 TPH Analysis by Infrared Spectrometry

Extracts were dried completely under a nitrogen stream. One gram of chromatographic silica gel (100- to 200-mesh chromatographic silica gel from Fisher Scientific activated at 110°C for 24 hr) and 10 ml Freon were added to analyze TPH. Adsorption was measured at 2930 cm^{-1} (HC-404 Buck Scientific, Inc., East Norwalk, CT). The infrared spectrometer was calibrated using an EPA Reference Oil Standard (Buck Scientific) containing isooctane, chlorobenzene, and hexadecane. A five-point standard curve was used to establish the relationship between absorbance and concentration. Absorbances of TPH in the Freon solutions were converted to concentrations of TPH in solution using a standard curve.

2.1.1.5 TPH Analysis by Gas Chromatography/Flame Ionization Detection

Although infrared (IR) spectrometry analysis is relatively simple and an adequate survey method, it is not sensitive to all components of petroleum hydrocarbons. Therefore, TPH was determined by gas chromatography (GC) as well as IR. Extracts were injected into a Hewlett-Packard 5890A gas chromatograph equipped with a 7673A autosampler (Hewlett-Packard, Avondale, PA). A DB-TPH capillary column with dimensions of 30 m × 0.32 mm internal diameter (ID) with a stationary-phase thickness of 0.25 mm (J&W Scientific, Folsom, CA) was used for analytical separation, and a flame ionization detector was used for analyte detection. Hewlett-Packard DOS Chemstation software was used for integrating the total area from 6 to 24 min of retention, which should encompass straight-chain hydrocarbons from C_{10} to C_{28} (diesel range) for Virginia samples. For California samples, the total area was integrated from 14 to 32 min of retention time, which covered straight-chain hydrocarbons from C_{16} to C_{34} for crude oil. Hydrogen was used for the carrier and fuel gas at flow rates of 5 and 45 ml/min, respectively. The support gas was zero-grade air delivered at 420 ml/min, and nitrogen was used as the make-up gas at 25 ml/min. The chromatography conditions were splitless injection mode, 2-ml injection volume, injector temperature of 250°C, and detector temperature 350°C. The initial oven temperature of 40°C was held for 2 min and was then increased at a rate of 12°C/min to a final temperature of 320°C. This temperature was held for 1 min for Virginia samples or 10 min for California samples. A 1.0-ml aliquot of each extract was transferred to a GC vial and spiked with 5-alpha androstane (Accustandard, Inc., New Haven, CT) as an internal standard. Analytical standards of diesel and motor oil were prepared over a range of 20 to 250 mg/l.

2.1.1.6 PAH Analysis by Gas Chromatography/Mass Spectrometry

PAHs extracted from the California oxidation pond sludge were analyzed by gas chromatography/mass spectrometry (GC/MS) after an initial silica gel column chromatography separation procedure. A chromatographic column with a Teflon stopcock (250 × 15 mm ID) was plugged with Pyrex glass wool at the end. Ten grams of silica gel (100- to 200-mesh chromatographic silica gel from Fisher Scientific, activated at 110°C for 24 hr) was mixed with 20 ml hexane and then transferred to the column. Another 20 ml of hexane was used to rinse the column. Twenty milliliters of each dichloromethane extract (from soil) was dried under a nitrogen stream, redissolved in 20 ml hexane, and loaded on the top of the column. The column was eluted with 120 ml hexane and 120 hexane and benzene (1:1/v:v). The eluent was dried under a nitrogen stream, redissolved in 1 ml dichloromethane, and transferred to a GC vial. A mixed internal standard (including acenaphthene-d12, chrysene-d12, naphthalene-d8, phenanthrene-d10; ULTRA Scientific, North Kingstown, RI) was added before GC/MS analysis. PAHs were analyzed by a Hewlett-Packard Model 6890 gas chromatograph equipped with a 7673A autosampler (Hewlett-Packard, Avondale, PA). An HP-5MS fused silica column with dimensions of 30 m × 0.25 mm ID (0.25-mm film) was used for analytical separation, and a mass selective detector (MSD) was used for analyte detection. The chromatography conditions were splitless injection mode, injector temperature of 290°C, 1-ml injection volume, and MSD temperature of 280°C. The oven temperature was held constant at 50°C for 1 min and then was increased at a rate of 6°C/min until the final temperature of 300°C was reached. For target analytes, selected ion monitoring was used for identification and quantification.

2.1.2 Results of Experiment 1: Extraction by Shaking with Different Solvents

The purpose of this experiment was to assess the ability of several solvents to extract PAHs from soil as a function of soil texture and soil moisture. For the coarse-textured Sarpy soil, the extraction of anthracene and pyrene was not impacted by soil moisture content except for the most nonpolar solvents (hexane and dichloromethane) at 100% saturation (Figure 2.1). At this high moisture content, the soil visibly repelled nonpolar solvents and formed clumps rather than dispersing, as observed with the other three solvents. For BaP, only acetone and methanol were able to extract more than 60% of the original 29 mg/kg.

Extraction of PAHs from the medium-textured Eudora soil was less efficient than from the Sarpy. Extraction of anthracene and pyrene from the Eudora soil was impacted very little by moisture content, except for hexane and dichloromethane at saturation and 50% saturation. However, air drying the soil significantly reduced extractability for both compounds by all solvents. Acetone removed nearly 100% of the BaP from moist soils (compared to a maximum of 80% for methanol), but even acetone extracted only 55% for the air-dried samples.

The trend toward decreased extractability with increasing clay content continued for the Smolan soil. Although acetone generally was the best extractant, efficiencies for all three compounds were unacceptably low (<70%) for air-dried soil regardless of the solvent used. For BaP, acetone extracted more than 75% of the original concentration for all but air-dried soil; none of the other solvents extracted as much as 45% of the BaP at any moisture content.

The choice of solvent is critical in developing a shaking extraction protocol. Strongly adsorbed compounds will be affected by the soil texture and moisture content. The results for the air-dried soils also suggest potential problems with aging. The moisture content of these soils was brought to 50% saturation after contamination, and then they were allowed to dry at ambient temperatures for a period of 72 hr. For all soils except the Sarpy, this process had a strong negative impact on extractability of all compounds. A more rigorous procedure will be needed to quantitatively remove these compounds.

2.1.3 Results of Experiment 2: Comparison of Shaking and Soxhlet Extractions

Based on the results from Experiment 1, acetone and dichloromethane (a typical solvent in EPA protocols) as extraction solvents were tested using a more rigorous shaking procedure. In this experiment, extractability of TPHs by shaking was determined as affected by soil weight, solvent, shaking time, and soil moisture and was compared to standard soxhlet extraction. The test materials were diesel-contaminated soil from Virginia and petroleum-contaminated sediments from California.

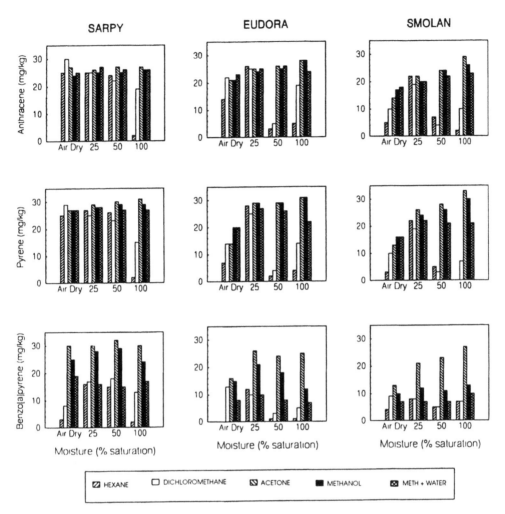

Figure 2.1 Extraction of anthracene, pyrene, and BaP as impacted by solvent, soil moisture, and soil texture.

2.1.3.1 Virginia Soil

Table 2.2 summarizes TPH concentrations determined in dry and moist Virginia soil with shaking and soxhlet extraction and quantified by IR spectrometry. For soxhlet extraction, five duplicate soil samples were analyzed with 2097 ± 50 mg TPH per kilogram for dry soil and 2172 ± 110 mg/kg for wet soil (mean $\pm$ confidence interval, $p < 0.05$). The mean soil concentrations for wet and dry soil were not significantly different, but the least significant difference (LSD) was greater for wet soil. Subsampling variability was higher in wet soil than dry soil, as a result of the greater degree of homogeneity of the dry soil that was ground and passed through a 40-mesh sieve. However, the fact that the mean TPH concentrations were the same for dry and wet soil indicated that moderate amounts of soil moisture did not interfere with soxhlet extraction of TPH with dichloromethane.

For shaking extraction, all factors were important, and only a few combinations resulted in TPH concentrations equivalent to soxhlet. For dichloromethane, only 1 g of soil in combination with three or four shaking cycles yielded TPH concentrations equivalent to

Table 2.2 Concentration of TPH (mg/kg) Determined in Diesel-Contaminated Soil from Virginia Using Different Solvents and Number of Sequential Extraction Cycles

Soil Mass	Time (hr)	Dichloromethane (cycles)				Acetone (cycles)				Dichloromethane + Acetone (cycles)			
		1	2	3	4	1	2	3	4	1	2	3	4
Air dry													
1 g	0.5	1701	1955	2035	2057	1828	2033	2066	2076	1771	2020	2054	2041
3 g	0.5	1536	1856	1825	1909	1512	1747	1849	1855	1331	1600	1726	1912
5 g	0.5	1427	1729	1879	1923	733	836	1088	1786	754	1583	1606	1828
1 g	1.0			2200	2223								
1 g	1.5			2207	2197								
3 g	1.0			2127	2161								
3 g	1.5			2125	2190								
5 g	1.0			1888	2113								
5 g	1.5			1978	1984								
Moist													
1 g	0.5	1494	1588	2113	2223	1583	2030	2187	2201	1961	2166	2246	2259
3 g	0.5	819	870	1100	1384	1328	1908	1882	2005	1579	1941	1838	1903
5 g	0.5	713	927	978	1246	1230	1581	1635	1882	1247	1847	1826	1870

Note: All values in the table are means of four replicates. LSD = 133 mg/kg for dry soil and 174 mg/kg for wet soil. Using soxhlet extraction, the TPH concentration was 2097 ± 50 mg/kg in dry soil and 2172 ± 110 in moist soil (mean of five replicates $\pm$ 95% confidence interval).

soxhlet for both wet and dry soils (Table 2.2). When 3 or 5 g of soil or fewer than three cycles was used, concentrations from shaking were significantly less than from soxhlet. When acetone was part of the solvent (as either 100 or 50% acetone), 1 g of soil in combination with two or more shaking cycles yielded TPH concentrations equivalent to soxhlet (Table 2.2). Greater amounts of soil or only one shaking cycle yielded significantly lower concentrations.

The trends in TPH concentration extracted by acetone-containing solvents provided an interesting contrast to dichloromethane alone for both wet and dry samples. For a given number of cycles with 3 or 5 g of wet soil, acetone and 1:1 acetone:dichloromethane extracted more TPH from soil than dichloromethane alone. For dry soil, the three solvents extracted equivalent concentrations for all cycles with 1 or 3 g, but with 5 g of soil, dichloromethane removed more TPH than the other two solvents for all numbers of cycles. The amount of water in 3 and 5 g of moist soil apparently was more than the dichloromethane could dissolve or displace, and the nonpolar solvent was unable to fully interact with the soil and remove the entrained hydrocarbons. Because acetone has both nonpolar and polar properties, its presence in the extraction solvent overcame the hydrophobicity that dichloromethane alone could not.

Increasing the extraction time per cycle had no impact on the TPH concentrations removed from the Virginia soil. Using 1 and 1.5 hr/cycle for three and four cycles, the concentrations did not significantly increase from the 2097 mg/kg removed by soxhlet.

2.1.3.2 California Soil

For the California material, only dry soil was used because of the ease of handling, lack of volatile compounds in the original samples, and the positive results from Experiment 1. Soxhlet extraction removed 7984 ± 90 mg/kg TPH as determined by IR analysis (Table 2.3). GC analysis of the soxhlet extracts clearly demonstrated that the hydrocarbons in this contaminated soil were associated with much longer carbon chains (up to at least C_{36}). This information, coupled with the age of the material and the results from the Virginia soil, led to the examination of only three and four extraction cycles. Fewer cycles were likely to be

Table 2.3 Mean TPH Concentration (mg/kg) in Petroleum-Contaminated Soils from California Using Soxhlet Extraction

Extraction Time (hr)	Dichloromethane		Acetone		Dichloromethane + Acetone	
	3 cycles	4 cycles	3 cycles	4 cycles	3 cycles	4 cycles
0.5	7993	8231	8792	8348	7853	8620
1.5	7853	8058	7826	8349	7959	7660
4.0	7359	8637	9012	8329	8705	8180

Note: 95% confidence interval ($p < 0.05$) was 7984 ± 90 mg/kg, LSD ($p < 0.05$) was 705 mg/kg, $n = 4$.

less efficient than soxhlet. The same solvents (dichloromethane, acetone, and 1:1 acetone:dichloromethane) and different extraction periods (0.5, 1.5, and 4.0 hr) were tested.

The shaking process extracted TPH concentrations that were equal to or greater than those from soxhlet for all conditions tested (Table 2.3). The variability was high for these samples, but the trends are readily apparent. Solvent, number of cycles, and extraction method did not have consistent effects on TPH concentrations extracted.

Despite different contaminants in the California and Virginia soils, shaking a 1-g soil sample with three successive aliquots of 10 ml of dichloromethane or acetone extracted concentrations of TPH equivalent to standard soxhlet extraction for both soils.

2.1.3.3 Extended Comparison for Virginia Soils

To compare soxhlet vs. shaking for a large number of samples, 96 samples were taken from the Virginia site in October 1995 and extracted with dichloromethane using 1 g dry soil and three shaking cycles. The TPH concentrations (determined by IR) in samples extracted by shaking were very similar to TPH from soxhlet with $R^2 = 0.82$ (Figure 2.2). The slope was statistically equivalent to 1.0, and the intercept was not significantly different from zero ($p < 0.05$).

An important aspect of this comparison is the reduction in labor required to obtain the extracts. Twenty-three hours was required to extract the 96 samples by shaking, in contrast to 72 hr required by soxhlet. This is a very significant savings in time without any loss in extraction efficiency.

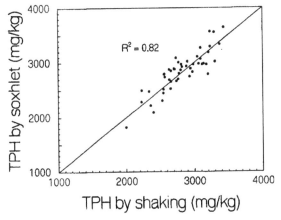

Figure 2.2 Comparison of TPH concentrations extracted from diesel-fuel-contaminated Virginia soil by shaking vs. standard soxhlet (evaluated by IR spectrometry).

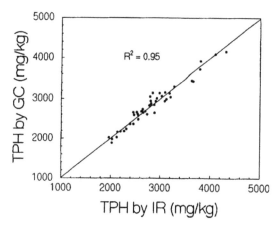

Figure 2.3   Comparison of TPH extracted from Virginia soil contaminated with diesel fuel as quantified by IR spectrometry and GC.

To evaluate TPH concentrations more accurately, GC also was used to analyze for TPH after shaking extraction. Concentrations of TPH from shaking extraction (1 g dry soil, three extraction cycles of 0.5 hr each) and quantified by IR were highly correlated ($R^2 = 0.94$) with TPH quantified by GC (Figure 2.3). Slope of the regression line was not significantly different than 1.0, and the intercept was statistically equal to 0.0. Thus, IR provided TPH concentrations equivalent to those determined by GC analysis, despite known shortcomings of the IR method (George, 1994).

2.1.4   Results of Experiment 3: Shaking vs. Soxhlet for Aged Contaminants

The objective of this experiment was to test extraction by shaking with either acetone or dichloromethane against soxhlet extraction for a suite of 17 compounds that had been added to soil and aged up to 1 year. Results for six representative compounds are discussed. All soil samples were extracted moist (approximately 10% H_2O). The initial concentrations were approximately 150 mg/kg soil for all compounds except pristane, which had an initial concentration of approximately 125 mg/kg.

A set of soils was amended with the contaminants immediately prior to extraction to serve as a measure of recovery by all methods. Recoveries of all compounds by all solvents and methods were essentially 100% (Table 2.4), with concentrations statistically equal to the known initial concentrations, except for a single cycle of shaking, which consistently extracted significantly less than 100% of the initial concentrations.

Aging had a strong effect on extractable concentrations by all methods. For samples aged for 48 days at 25°C followed by 180 days at 4°C, shaking with either acetone or dichloromethane for three cycles extracted approximately 75% of the original levels (Table 2.5). The concentrations were probably reduced by volatilization and degradation during aging before extraction began. Soxhlet extraction after grinding with Na_2SO_4 to eliminate moisture removed significantly lower concentrations of all analytes except pyrene when compared to three shaking cycles with acetone and significantly lower concentrations of tetradecane, pristane, 1-octadecene, and phenanthrene when compared to three shaking cycles with dichloromethane.

Aging for 180 days outdoors followed by 180 days at 4°C further decreased extractable concentrations (Table 2.6). Shaking with acetone was the most effective method. Recovery

Table 2.4 Extraction of Compounds from Simulated Diesel/Soil Mixture Using Soxhlet, Shaking with Acetone, or Shaking with Dichloromethane as the Extraction Method

Compound	Acetone Shaking Cycles			Dichloromethane Shaking Cycles			Soxhlet + Na_2SO_4	LSD $p < 0.05$
	1	2	3	1	2	3		
Tetradecane	137	142	141	135	135	150	152	9
Pristane	117	123	123	122	123	124	124	12
1-Octadecene	139	146	145	140	147	147	148	9
Phenanthrene	131	136	137	142	142	141	133	15
Pyrene	133	139	139	143	143	141	134	14
Tetracosane	135	143	140	139	137	138	136	14

Note: 150 mg/kg of each compound (125 mg/kg for pristane) was added.

by soxhlet often was <50% of that by shaking with acetone. Shaking with dichloromethane was generally the least effective extraction method except in the case of pyrene.

2.1.5 Results of Experiment 4: PAHs Extracted by Shaking vs. Soxhlet

GC/MS was used to quantify individual PAHs extracted using the shaking and soxhlet methods. Both extraction methods were performed in quadruplicate on a single sample of diesel-contaminated Virginia soil. Shaking and soxhlet extracted the same pattern of PAHs (Figure 2.4), and the mean concentrations of each PAH extracted by shaking were equal to or greater than concentrations obtained by soxhlet ($p < 0.05$).

The shaking and soxhlet extraction methods were compared on a wide range of contaminated soils. A single cycle of shaking was adequate for complete recovery of anthracene and pyrene from a moist, sandy soil with water content ranging from air dry to 100% of saturation. The low-polarity solvents hexane and dichloromethane had very poor recoveries of BaP at all moisture contents, but acetone maintained complete recovery. Extraction efficiencies for all solvents declined with increasing clay content, increasing PAH molecular weight, and after air drying contaminated soil initially at 50% moisture saturation. Low recoveries probably occurred because the solvents were unable to remove PAHs that were strongly sorbed or entrained in the soil micropores during a single short-duration shaking extraction. Under all circumstances, acetone consistently was the best solvent for extraction.

In an effort to improve shaking-extraction efficiency, the effects of drying, grinding, and multiple sequential extractions were tested. For aged, contaminated soils from sites in Vir-

Table 2.5 Extraction of Simulated Diesel Fuel Components Added to Eudora Soil, Then Aged for 48 Days at 25°C and 180 Days at 4°C Under Moist Conditions

Compound	Acetone Shaking Cycles			Dichloromethane Shaking Cycles			Soxhlet + Na_2SO_4	LSD $p < 0.05$
	1	2	3	1	2	3		
Tetradecane	102	102	107	104	109	120	86	11
Pristane	92	93	98	85	91	100	82	11
1-Octadecene	109	110	114	99	105	116	93	12
Phenanthrene	100	101	106	92	100	108	93	12
Pyrene	102	103	107	91	101	108	96	13
Tetracosane	105	107	112	93	98	105	92	14

Note: 150 mg/kg of each compound (125 mg/kg for pristane) was added.

Table 2.6 Concentration of Simulated Diesel Fuel Components Added to Eudora Soil, Then Aged for 180 Days at 25°C Followed by 180 Days at 4°C Under Moist Conditions

Compound	Acetone Shaking Cycles			Dichloromethane Shaking Cycles			Soxhlet + Na$_2$SO$_4$	LSD $p < 0.05$
	1	2	3	1	2	3		
Tetradecane	18	20	23	4	5	nd[a]	13	3
Pristane	56	61	68	20	24	24	35	6
1-Octadecene	46	50	55	13	13	15	28	6
Phenanthrene	4	8	11	2	nd	5	nd	6
Pyrene	13	17	9	20	14	12	13	11
Tetracosane	74	80	81	19	24	21	41	7

Note: 150 mg/kg of each compound (125 mg/kg for pristane) was added.
[a] nd = not detected.

ginia and California, three sequential extractions by shaking with acetone or dichloromethane for 30 min recovered TPH concentrations that were equivalent to soxhlet extraction. This same shaking procedure extracted 13 PAHs from the California oxidation pond sludge at concentrations that were equal to or greater than soxhlet extraction. Drying and grinding had no impact on TPH concentrations recovered by shaking when 1 g of contaminated soil was extracted; however, when 3 g or more of soil was used, extraction efficiencies generally were <100% regardless of the solvent used or if the soil was extracted moist or dry.

From these results, either acetone or dichloromethane can be used to extract 1 g of moist or air-dried soil if three or more sequences of shaking are used. For field-moist soils, acetone consistently worked better than dichloromethane if 3 g or more of moist soil was extracted. The method chosen for the Virginia samples was three sequences of extraction of 1 g of dried, ground soil with dichloromethane. The results provided confidence that this method should yield results at least as high as soxhlet extraction, and the drying and grinding should minimize subsampling errors.

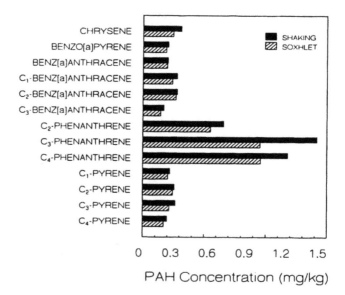

Figure 2.4 Concentrations of a suite of PAHs extracted from Virginia soil by shaking and soxhlet and quantified by MS.

2.2 DISSIPATION OF BENZO[A]PYRENE IN
THE RHIZOSPHERE OF *FESTUCA ARUNDINACEA*

PAHs contain two or more benzene rings in a linear, cluster, or angular arrangement in which adjacent rings share two carbons. Microbial degradation and transformation are the primary processes that remove PAHs from the environment. However, four- and five-ring PAHs are recalcitrant because of the resonance energies of their structures and their low water solubilities (Klevens, 1950; Heitkamp and Cerniglia, 1988).

The presence of organic matter significantly enhances degradation of PAHs. Mahro et al. (1994) reported accelerated dissipation of PAHs after addition of compost. They noted increased mineralization and formation of nonextractable bound residues and hypothesized that the formation of bound residues represents an important depletion mechanism in soil. Metabolic by-products that are formed from PAH biodegradation (e.g., hydroxy carbonic acids or phenolic compounds) may interact chemically with soil organic compounds, be incorporated into the humus, and thereby form bound residues.

Bacteria and fungi in the rhizosphere have the potential to biodegrade PAHs aerobically. Bacteria produce dioxygenases that incorporate two oxygen atoms into PAHs to form dioxyethanes which are further oxidized to dihydroxy by-products (Gibson et al., 1975; Cerniglia and Heitkamp, 1989). Fungi produce monooxygenases that incorporate one oxygen atom into the PAH to form arene oxides. The metabolites are further degraded to produce succinic, fumaric, pyruvic, and acetic acids and acetaldehyde, which are used for protein synthesis (Sims and Overcash, 1983). PAHs can be degraded either as a source of carbon and energy or by cometabolism. Bacterial degradation of PAHs with four or more rings by cometabolism has been observed (Heitkamp and Cerniglia, 1988).

The use of plants to enhance bioremediation of contaminated soil has been described as an effective cleanup method (Walton and Anderson, 1990, 1992; Anderson et al., 1993). Evidence exists for the enhanced dissipation of PAHs, insecticides, and herbicides in the rhizosphere when compared to nonrhizosphere soil (Hsu and Bartha, 1979; Reddy and Sethunanthan, 1983; Sandmann and Loos, 1984; Lappin et al., 1985; Aprill and Sims, 1990; Ferro et al., 1994; Gunther et al., 1996).

The bioavailability of highly recalcitrant, carcinogenic PAHs is of concern for accurate risk assessment. However, the ultimate fate of contaminants in vegetated soil is unclear. The objective of this study was to investigate the fate of ^{14}C-labeled BaP in the rhizosphere of tall fescue (*Festuca arundinacea*) by evaluating the distribution of ^{14}C in the soil, plant tissue, and gas phase and the residual concentration of the parent compound.

2.2.1 Materials and Methods

A 6-month greenhouse study evaluated the degradation of ^{14}C-BaP and the ultimate fate of ^{14}C in vegetated and unvegetated systems. Plants growing in ^{14}C-BaP-contaminated soils were contained in large Plexiglas chambers to allow for monitoring of ^{14}CO$_2$, ^{14}C volatile organics, plant uptake, parent compound dissipation, and final distribution of the ^{14}C in soil.

2.2.1.1 Soil Treatments

A Eudora silt loam soil (coarse-silty, mixed, mesic Fluventic Hapludolls) was collected from Kansas State University Ashland Agronomy Farm near Manhattan, KS. The soil had a pH of 6.1 (1:1 H$_2$O:soil); 2.0 mg/kg of exchangeable NH$_4$–N; 10 mg/kg of soluble NO$_3$–N; 67 mg/kg as K (1 M NH$_4$OAc); 0.3% organic C; a cation exchange capacity of 49 mmol/kg

(NH_4OAc, pH 7); and 41% sand, 45% silt, and 14% clay (hydrometer method). Each chamber was packed with 1.2 kg of soil (<2 mm) that had been amended with BaP as follows: 13.16 μCi of ^{14}C-BaP (carrier free, 98% purity, Sigma Chemical Company, St. Louis, MO) was dissolved in acetone along with unlabeled BaP to achieve a total of 60 mg BaP. The acetone was sprayed onto 1.2 kg soil in three equal increments. The soil was mixed thoroughly between each increment. The acetone was allowed to evaporate completely before packing the soil into the chambers.

2.2.1.2 Plant Establishment

Tall fescue (*Festuca arundinacea*), a cool-season grass, was selected for the experiment. This grass has an extensive rooting system and is often used for soil stabilization. Triplicate plant chambers were used for each treatment. Seedlings were placed on circular wax plugs (1 cm thick × 6 cm in diameter). The wax mixture (two parts petroleum jelly and one part paraffin) allows penetration by the roots but restricts air transfer. A 1-cm layer of vermiculite was placed over the seeds and kept moist. One-month-old seedlings were transplanted into the growth chambers, which were then placed in a controlled-temperature (21°C) greenhouse. Water was added daily to adjust the soil to an appropriate moisture content (80% filled void space). Leachate production was minimal during the experiment and contained no detectable BaP. Commercial fertilizer was added to all treatments every month (MiracleGro at 10 lb/ acre). Natural light was used for this greenhouse study. After 6 months of plant growth, soil and plant samples were removed from the chambers for analysis.

2.2.1.3 Growth Chamber Design

Cylindrical, Plexiglas, plant growth chambers (Figure 2.5) were used for monitoring the fate of ^{14}C in the soil–plant system. The upper section of the chamber that contained the plant shoots was 40 cm in height and 20 cm in diameter. The lower section, containing plant roots, was 20 cm in height and 10 cm in diameter. A Plexiglas barrier separated the upper and lower sections with a central opening into the soil chamber to accommodate the wax plug, which was placed in the opening after seedling development. Air was continuously evacuated by vacuum through the upper section at a constant rate (40 ml/min) and passed through a series of traps. Simultaneously, air was evacuated through the lower section at a constant rate (10 ml/min) and passed through a similar set of traps. Air leaks and steady-state conditions were checked by flow meters located in the upstream and downstream sections of the chamber.

A series of four traps was used for sampling of the gas phase. Each trap was a 20-ml glass vial equipped with a two-hole stopper for inlet and outlet tubes. The first and last vials were empty to trap droplets carried by the air stream. The second vial was filled with 10 ml of Aquasol (Packard Instrument Company, Downer Grove, IL) to trap volatile ^{14}C organic compounds (Hsu and Bartha, 1979). The third vial contained 10 ml of Carbosorb II for $^{14}CO_2$. A charcoal filter was used as a final trap to capture any volatile ^{14}C compounds that may have escaped the Aquasol. The efficiency of this system (not including the charcoal) was greater than 99%. Trapping solutions were replaced daily. All samples were refrigerated until analysis. Samples were analyzed with a Tri-Carb 4000 Series liquid scintillation counter (Packard Instrument Company).

2.2.1.4 Soil Analysis

Soil was sampled from each chamber at the end of 6 months when the plant was removed. The soil sample was homogenized, placed in vials, and stored at 4°C until

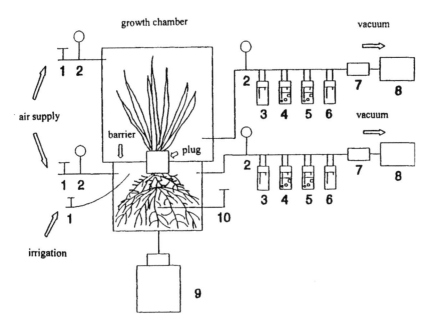

Figure 2.5   Design of plant growth chamber. 1 = valve, 2 = flow meter, 3 and 6 = empty vials, 4 = ^{14}C volatiles trap, 5 = $^{14}CO_2$ trap, 7 = activated carbon filter, 8 = vacuum pump, 9 = leachate collection, 10 = tensiometer.

analysis. The ^{14}C compounds in the soil subsamples (2 g) were oxidized to $^{14}CO_2$ by wet combustion (Allison, 1960). Several modifications were made to this method. The soil sample was digested in a 60:40 mixture of sulfuric acid and phosphoric acid in the presence of $K_2Cr_2O_7$. The boiling temperature of this mixture, 210°C, is high enough to ensure complete oxidation of carbonaceous matter, yet low enough to prevent excessive fuming in the condenser. This oxidation mixture is capable of attacking and dehydrating the resistant forms of carbon and reduces the boiling time for complete oxidation. Purification steps include a concentrated sulfuric acid trap to capture moisture and a U-tube containing granular zinc to absorb acid fumes. Anhydrous magnesium perchlorate was used to absorb water from the carrier stream containing evolved $^{14}CO_2$. The $^{14}CO_2$ in the carrier stream was trapped by two successive vials of Carbosorb II. Other than these modifications, the Allison method was followed for soil preparation and wet combustion of soil organic carbon to $^{14}CO_2$ (Allison, 1960). After combustion, samples were tightly capped and refrigerated at 4°C until analysis.

BaP was extracted from soil samples (2.5 g) by shaking for 60 min with 10 ml of acetone in 20-ml scintillation vials. The soil slurry was then filtered through Whatman #42 paper and the clear filtrate was collected. This method has a recovery of >95%. Approximately 1 ml of filtrate was transferred to a glass GC vial. All reusable glassware in the study was cleaned, rinsed in distilled deionized water, and baked at 280°C for 8 hr to remove organic residues.

Humic and fulvic acid fractions were quantified in the following manner. Thirty grams of soil was air dried. Three hundred milliliters of 0.5 N NaOH was added to each soil sample. The sample bottles were then purged with N_2, closed, and shaken for 24 hr at room temperature. Samples were centrifuged for 10 min at 2500 rpm to separate the dark-colored supernatant composed of humic and fulvic acids from the soil. After acidification to a pH of 2 with 2 N HCl, the soluble material (fulvic fraction) was separated from the insoluble humic fraction (Schnitzer, 1982). The ^{14}C in the soluble fulvic fraction was measured using liquid scintillation counting. The humic fraction was determined by vacuum drying the solid material and proceeding with the wet combustion method described earlier.

2.2.1.5 Plant Biomass Analysis

Uptake of ^{14}C into plant biomass was measured by removing plants from the chambers and separating roots from shoots. Shoots were cut at the soil level and pulverized with liquid N_2. Large roots were manually separated from soil, and small roots were separated by passing the soil through a 0.25-mm sieve. Both large and small roots then were washed and pulverized in liquid N_2. Triplicate plant subsamples (0.1 g) for each growth chamber were combusted by the wet combustion method. $^{14}CO_2$ was collected in a trapping unit and analyzed for radioactivity. Plant biomass also was measured on a dry weight basis for each plant chamber.

2.2.1.6 Microbial Enumeration

Soil microorganisms were enumerated by direct and plate counting methods and were replicated six times. One gram of soil from each soil treatment was placed in 10 ml of 0.85% sterile sodium chloride solution and then aseptically and serially diluted. For each dilution between 10^{-4} and 10^{-7}, 0.1 ml was plated in duplicate on peptone tryptone yeast extract agar medium (5 g peptone, 5 g tryptone, 10 g yeast extract, 10 g glucose, 0.5 g magnesium sulfate, 0.07 g calcium chloride, 15 g Difco Bacto agar, 1 l sterile, distilled, deionized water). Colonies were counted after 3 days of incubation at 25°C. Direct bacterial counting was done using epifluorescent microscopy. Total cell numbers were quantified using 4',6-diamidino-2-phenylindole (Yu et al., 1995).

2.2.1.7 Gas Chromatography

BaP was analyzed using a Hewlett-Packard 5890A gas chromatograph equipped with an HP3396 integrator, HP7673A autosampler, J&W DB-5 capillary column (J&W Scientific, Folsom, CA), and flame ionization detector (FID). After injection of the sample, an extended isothermal period (35°C for 6 min) was implemented to ensure the signal returned to near baseline after the large solvent peak tail. The oven temperature was programmed to increase rapidly at an initial rate of 70°C/min to 110°C, followed by a secondary heating rate of 20°C/min to a temperature of 300°C. The carrier gas and source of fuel for the FID was H_2, and N_2 served as the make-up gas.

2.2.2 Results and Discussion

After 185 days, the plants in the vegetated chambers had extended their leaves very near the top of the upper cylinder. The 40-cm height of the upper cylinder minimized crowding, and the plants were healthy throughout the experiment. The average dry biomass of the shoots and roots was 1.31 and 0.56 g, respectively. The mass of soil used (1.2 kg) was great enough to allow for root expansion throughout the soil without the plant becoming root bound. These data indicated that this plant species can survive in soil contaminated with the target compound. Previously reported research indicated that BaP may act as a growth stimulant at lower concentrations (10 mg/kg) (Graf, 1965), but may be phytotoxic at higher concentrations (3.2 mg/kg) (Forrest et al., 1989). As an uncontaminated control was not included in the present study, the effect of the contaminant on plant growth cannot be evaluated. In both the

Table 2.7 Percent Distribution of ^{14}C After 185 Days of Plant Growth

	%		
	With Plants	**Without Plants**	**LSD ($p < 0.05$)**
$^{14}CO_2$			
Upper (shoot) section	0.23	—	—
Lower (root) section	1.05	—	—
Total	1.28	0.66	0.07
Volatile organics	0.15	0.03	0.06
Plant biomass			
Roots	0.12	—	—
Shoots	0.01	—	—
Soil	92.32	95.46	ns[a]
Total recovery	93.88	96.15	—

[a] ns = not significant.

vegetated and unvegetated chambers, BaP had been transformed by the end of the experiment (Table 2.7). The mechanisms of dissipation evaluated were mineralization, plant uptake, and incorporation of the contaminant into soil organic material. Photodegradation and volatilization, which can contribute to loss of some PAHs, are not significant pathways of dissipation for BaP in soil (Sims and Overcash, 1983; Park et al., 1990).

Mineralization of ^{14}C to $^{14}CO_2$ accounted for only 1.28% of the labeled ^{14}C in the planted chambers by the end of the experiment. Approximately half of this amount (0.66%) was mineralized in the unplanted control (Figure 2.6). Differences in cumulative mineralization between the planted and unplanted soils were not significant until approximately 30 days, which may be the time needed for establishment of a substantial rhizosphere. Cumulative $^{14}CO_2$ was four times greater in the soil chamber compared to the plant shoot chamber. The $^{14}CO_2$ was higher than expected in the upper plant shoot chamber and, given the very low amount of ^{14}C found in plant shoot biomass, may be due to a small leak in the wax plug through which plant roots were growing. Since the ^{14}C-BaP had only a 98% purity rating, it is not possible to conclude that any mineralization of the parent compound occurred. However, degradation of BaP to CO_2 was not expected to be a significant pathway for dissipation in this study. Complete bacterial mineralization of PAHs has been limited primarily to compounds with two, three, and four aromatic rings (Heitkamp and Cerniglia, 1988; Mueller et al, 1991; Walter et al., 1991; Weissenfels et al., 1991).

Volatilization of ^{14}C compounds other than $^{14}CO_2$ was slightly higher in the planted soil chambers when compared to unplanted chambers but accounted for only 0.15% of the total ^{14}C added to the system. BaP is not volatile in soil (Sims and Overcash, 1983); therefore, volatile intermediate metabolic products probably contributed to this fraction.

Plant uptake of ^{14}C was minimal in this study. The data showed that very little ^{14}C label was translocated into plant shoots, and slightly more ^{14}C was present in the plant roots. Plant tissue concentration and distribution patterns of ^{14}C were as expected given the strong adsorption characteristics and low aqueous solubility of BaP. These results also indicated that metabolic by-products present in the soil were not readily taken up in the transpiration stream. These findings are consistent with previously reported results. Previous studies have found that the concentration of BaP in upper plant parts was not a function of soil concentration (Shabad and Cohan, 1972), and translocation of BaP did not occur in several species of plants (Blum and Swarbrick, 1977). The concentration of BaP in the outer layer of the plant roots, however, has been shown to correlate with soil concentration (Borneff et al., 1973).

Soil microbial numbers were higher in the planted soil than unplanted soil (Table 2.8). The planted soil had an order of magnitude more bacteria as determined by both plate and

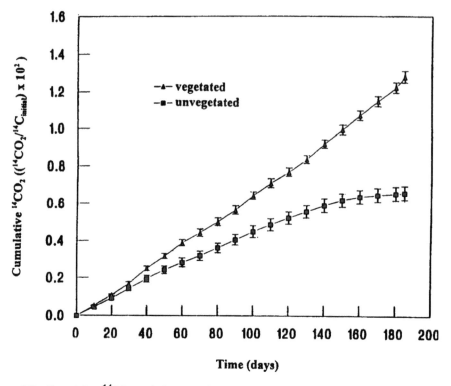

Figure 2.6 Cumulative $^{14}CO_2$ evolution over the experimental period.

direct counts. The increase in microbial numbers was expected because soil adjacent to roots usually contains increased microbial numbers and populations (Paul and Clark, 1989). The number of bacteria in the rhizosphere can be significantly higher than that normally found in nonrhizosphere soil (Rovira and Davey, 1974), even in contaminated soils (Lee and Banks, 1993). The increased microbial numbers are primarily due to the presence of plant exudates and sloughed biomass that serve as sources of energy, carbon, nitrogen, or growth factors.

The residual BaP concentration was significantly ($p < 0.05$) less in planted soil than in unplanted soil (Table 2.8). Fifty-six percent of the applied BaP was degraded in vegetated soil compared with 47% in the unvegetated soil. Similar fractions of BaP degradation (72% to 33%) in soil were observed by Khesina et al. (1969) in a 3-month study. Also, Mueller et al. (1991) reported 43% degradation of BaP in soil after 12 weeks of tillage and fertilization. By directly comparing vegetated and unvegetated contaminated soils, results from this study indicate that contaminant dissipation can be enhanced by the presence of plant roots.

Table 2.8 Comparisons of Residual Contaminant Concentrations and Microbial Numbers in Planted and Unvegetated Soil

	With Plants	Without Plants	LSD ($p < 0.05$)
Microbial enumeration			
Plate counts (colony-forming units × 10^6/g dry soil)	4.2	0.42	0.32
Direct counts (# × 10^5/g dry soil)	3.0	0.19	0.29
Residual contaminant concentration			
BaP (mg/kg [% C/C_0])	22.1 (44.2%)	26.5 (53.0%)	2.15 (4.3%)

The primary sink for the ^{14}C label in the study was the soil matrix. Ninety-two percent of the ^{14}C was found in the vegetated soil and 95% in the unvegetated soil. The rates and extent of humification in planted soil are greater than in unplanted soil (Stevenson, 1982). Humification of toxic contaminants may decrease bioavailability and, consequently, reduce toxicity (Walton et al., 1994). To determine if humification of ^{14}C-BaP was a significant mechanism of dissipation in this study, ^{14}C in the fulvic/humic acid fractions was quantified. In the unplanted soil, 11.1% of the ^{14}C label was associated with the combined humic and fulvic acid fractions, as compared to 10.9% in the vegetated soil. This small difference was not statistically significant. Walton et al. (1994) also observed no statistical difference between the percentage of ^{14}C-pyrene associated with fulvic/humic acid fractions in soil vegetated with white sweet clover (*Melilotus alba*) (4.3%) and unvegetated soil (4.4%) after 5 days. Differences were noted, however, for two- and three-ring PAHs (naphthalene and phenanthrene). Although a significant portion of ^{14}C (approximately 11%) found in the soil after this 6-month study was associated with the fulvic and humic fractions, humification of BaP was not accelerated in the rhizosphere. A longer experimental period may be required to observe significant differences in the extent of humification for the more recalcitrant four- and five-ring PAHs.

2.2.3 Conclusions

The major sink for ^{14}C-BaP in a plant–soil system was the soil matrix in both planted and unplanted soil. Although more than half of the parent contaminant degraded in vegetated soil during the 6-month experimental period, over 90% of the ^{14}C label remained in the soil. Microbial numbers were significantly higher in planted soils when compared to unvegetated soils. Significantly more degradation of BaP occurred in the planted soil when compared to the unvegetated soil, indicating that the rhizosphere enhanced contaminant degradation. However, the extent of humification of ^{14}C-BaP in the planted soil was not found to be significantly higher than in unplanted soils. Mineralization, volatilization, and plant uptake of BaP were found to be minor pathways of dissipation. There was no evidence of phytotoxicity from BaP.

In the plant–soil environment, BaP is degraded and by-products of degradation ultimately become incorporated into the soil matrix. Degradation occurs at a faster rate in the presence of plants, and degradation by-products are not readily available for mineralization by microbes or uptake by plants. The results from this study indicate that plants can significantly influence the fate of BaP in soil.

2.3 EFFECT OF SOIL DEPTH AND ROOT SURFACE AREA ON PHYTOREMEDIATION EFFICIENCY

Biodegradation of organic contaminants in soil has been observed to decrease with increasing soil depth. Research on pesticides as well as industrial pollutants has revealed that soil depth is a major consideration in the effectiveness of *in situ* bioremediation (Veeh et al., 1996; Strand et al., 1995). The effects of vegetation on soils contaminated with heavy metals also decrease with depth (Ou et al., 1995). The relationship between degradation and soil depth has not yet been assessed for plant-enhanced bioremediation of organic contaminants. Root morphology may play a role in the effectiveness of phytoremediation. Plants with a fibrous root structure, and therefore greater root surface area, may enhance organic dissipation more than plants with simpler, less fibrous systems (Aprill and Sims, 1990).

This study was designed to examine the effects of rooting depth and root surface area on the biodegradation of petroleum hydrocarbons in soil. The objectives were to (1) compare the degradation of anthracene, benzo[a]anthracene (BA), BaP, and TPH (diesel range) at three different soil depths in soil columns with vegetated and unvegetated soils; (2) monitor the microbiological activity in soils and compare microbial populations at all three soil depths for vegetated, unvegetated, contaminated, and uncontaminated soils; (3) quantify the growth of the root system in the soil; and (4) evaluate relationships with the biodegradation of organic contaminants.

2.3.1 Materials and Methods

The soil used in this experiment was a Haney sandy loam from the north agricultural farm, Department of Agronomy, Kansas State University, Manhattan, KS. Prior to use, the soil was analyzed for pH, phosphorus (P), potassium (K), ammonium nitrogen (NH_4^+–N), nitrate nitrogen (NO_3^-–N), organic matter content, cation exchange capacity (CEC), texture analysis, and background levels of anthracene, BA, and BaP (Table 2.9). The soil was passed through a No. 8 sieve to ensure uniformity prior to use.

The soil was contaminated with a mixture of the three target PAH compounds, plus commercial diesel fuel to represent contaminated field conditions. Anthracene was obtained from AKCROS Chemicals, Inc. (New Brunswick, NJ). BA was obtained from Sigma Chemical Company (St. Louis, MO). BaP was also obtained from Sigma Chemical Company. The PAHs and diesel fuel were dissolved in ACS-grade acetone (Fisher Scientific, St. Louis, MO). The solution was added to dry soil at 500 ml/10 kg-batches of soil. The final concentration of contaminants in the soil was 75 ppm anthracene, 50 ppm benzanthracene, 30 ppm BaP, and 2000 ppm diesel fuel by weight. The solution was applied to the soil in 50-ml increments, while the soil was continuously mixed in a large mechanical mixer. After application, the soil was placed in 20-gal plastic-lined buckets and allowed to set for 48 hr. Every 12 hr, the soil was turned to allow volatile compounds to escape prior to filling the columns. A 1-kg sample of the contaminated soil was placed in a sealed plastic bag and left in storage at 4°C for the duration of the experiment as an abiotic control. Plastic columns 30 in. tall were filled with 10 kg of soil each, resulting in a soil depth of approximately 26 in.

Table 2.9 Characterization of Haney Sandy Loam Used in Greenhouse Study

Parameter	Value
pH	7.9
CEC	8.2 meq/100 g
Moisture	1.48%
Organic matter	1.4%
Phosphorus	42 mg/kg
Potassium	205 mg/kg
NH_4^+–N	3.1 mg/kg
NO_3^-–N	10.8 mg/kg
Sand	47%
Silt	44%
Clay	9%
Anthracene	nd[a]
BA	nd
BaP	nd
TPH	nd

[a] nd = not detected.

After being saturated with tap water and allowed to settle, the depth of each column was approximately 24 in.

Switchgrass seedlings (*Panicum virgatum*) were transplanted to the saturated soil columns following germination and 2 weeks in a growth chamber at 80°F day/60°F night with a 16-hr photoperiod. Each vegetated pot received four plants. Thirty-six pots were assembled, with nine pots containing vegetation and contaminated soil (VC), nine containing vegetation and clean soil (VU), nine with no vegetation and contaminated soil (UC), and nine with no vegetation and clean soil (UU). All 36 pots were kept in a greenhouse at 70°F day/55°F night with a 15-hr photoperiod. The soil was watered as needed based on visual inspection (approximately 100 ml every other day). Leachate was collected in a pan and analyzed as described below. Fertilizer was applied every 3 weeks to provide a total C:N:P ratio of 100:20:5 based on a C measurement of 2000 ppm provided by the diesel fuel. Urea (46% N) [(NH$_2$)$_2$CO] was applied at a rate of 1.0 g per pot, and ammonium phosphate [(NH$_4$)$_2$HPO$_4$] was applied at 0.64 g per pot.

A time 0 soil analysis was performed immediately after transplanting the switchgrass seedlings. Column takedown and destruction occurred every 7 weeks over a 21-week period. For each takedown, three columns of each treatment (VC, VU, UC, UU) were cut into three 8-in. sections for separate analysis of top, middle, and bottom soil and biomass layers. Plant material was separated from the soil at each layer and kept at 4°C storage until analysis. Each soil layer was homogenized separately and kept in cold storage.

2.3.1.1 Standard Soil Extraction Procedure

The following procedure was used to extract and analyze PAHs and TPH from soil samples if the soil extracts did not require concentration. Three grams of wet soil sample was placed in a 20-ml scintillation vial along with 10 g of ACS-grade acetone (Fisher Scientific, St. Louis, MO). One hundred microliters of 1000-ppm tetracosane was added as the matrix spike component for quality assurance/quality control. The vial was placed in a reciprocal shaker and shaken for 30 min and was then centrifuged for 10 min at 5000 rpm. The supernatant fluid was decanted from the vial into a 60-ml glass bottle and tightly capped. This process was repeated with the original 3 g of soil sample (total of two extraction sequences). Approximately 1.6 ml of the extract was transferred into a 2-ml glass GC vial along with 5 ml of androstane as the internal standard. The vials were closed using crimp-top caps lined with a Teflon septum.

2.3.1.2 Soil Extract Concentration

The following procedure was used to concentrate soil extracts when standard extraction and analysis proved insufficiently sensitive to be used. A 10-ml aliquot of the soil extract was transferred into a 20-ml scintillation vial. The vial was placed under a fume hood and exposed to a stream of nitrogen gas (N$_2$) until the acetone had completely evaporated (approximately 1.25 hr). One milliliter of Optima Grade acetone (Fisher Scientific, St. Louis, MO) was then added to the vial. The vial was capped and shaken by hand to redissolve the deposited contaminants into the solvent. The 1-ml solution was transferred into a 2-ml glass GC vial along with 5 ml of androstane as the internal standard. The vials were closed using crimp-top caps lined with a Teflon septum.

All soil extract samples from these procedures were analyzed by GC. All glassware used in this procedure was previously washed with soap and tap water, rinsed twice with deionized water, and baked at 300°C for a minimum of 8 hr.

2.3.1.3 Leachate Analysis

The following procedure was used to extract PAHs and TPH from collected leachate. A 30-ml sample was used. If less than 30 ml of leachate was collected, the sample was mixed with distilled, deionized water to create 30 ml of solution. SEP-PAK C_{18} cartridges were conditioned sequentially with 5 ml of dichloromethane, 5 ml of methanol, and 5 ml of distilled, deionized water. The leachate sample flowed through the C_{18} cartridge by suction. PAHs and TPH were eluted using 3 ml of dichloromethane, dried with sodium sulfate, and transferred to 2-ml GC vials. A 5-ml volume of androstane was added to the vial as an internal standard.

All leachate samples were analyzed by GC as described below. All glassware used in this procedure was washed with soap and tap water, rinsed twice with deionized water, and baked at 300°C for a minimum of 8 hr.

2.3.1.4 Plant Extraction

The following procedure, adapted from Al-Assi (1993), was used to extract and analyze TPH and PAHs in plant tissue after sampling at 21 weeks. Up to 10 g (dry weight) of plant tissue was ground using a mechanical plant grinder. If 10 g was not available, the entire sample was used. The ground plant material was placed in a cellulose thimble, covered with a piece of filter paper, and extracted in a soxhlet apparatus using approximately 100 ml of methanol as a solvent. Two hundred microliters of tetracosane was added to the thimble as a component of quality assurance/quality control. The soxhlet was allowed to cycle for 3.5 hr and then allowed to cool. The methanol extract was transferred to clean 160-ml bottles and stored at 4°C until elution through C_{18} cartridges. SEP-PAK C_{18} cartridges were conditioned sequentially as follows: 5 ml of methanol, 10 ml of hexane, 5 ml of methanol, 5 ml of 10-ml methanol/100-ml H_2O. All solvents were gravity fed through a C_{18} cartridge.

The plant extract was then passed through the C_{18} cartridge and allowed to flow completely out using vacuum suction. The contaminants were eluted with 2.5 ml of hexane, dried with Na_2SO_4, and transferred to GC vials. The vials were sealed with crimp-top caps.

2.3.1.5 Gas Chromatography

Analysis of all extraction samples was performed using a Hewlett-Packard 5890A gas chromatograph equipped with an HP3396 integrator, HP7673A autosampler, J&W DB-5 capillary column (J&W Scientific, Folsom, CA), and FID. The carrier gas and fuel source for the FID was H_2. The flow rate for the carrier gas was 4.5 ml/min, while the fuel source was supplied at 45 ml/min. Nitrogen was supplied as the make-up gas at a rate of 25.5 ml/min. Air was supplied as the oxidant at a rate of 420 ml/min. The 40°C initial oven temperature of the gas chromatograph was maintained for 2 min. The temperature was then increased at a rate of 12°C/min until reaching a maximum of 320°C. The temperature was held for 1 min to complete the heating cycle. The temperature of the injection port was set at 250°C, while the detector temperature was 350°C. The volume of the sample was 2 ml.

Individual chromatograph peak areas for the PAH compounds were converted into concentrations using standard curves. TPH concentrations were measured using integrated GC areas covering a range of 5.9 to 24 min. These areas were also converted into concentrations using standard curves. New sets of standards were prepared for each of the sampling times. Table 2.10 provides concentration ranges for all three compounds in the standards. All standards were prepared using PAH compounds and diesel fuel dissolved in ACS-grade acetone (Fisher Scientific, St. Louis, MO).

Table 2.10 Concentration Ranges for GC Standards

Compound	Concentration Range (ppm)	Extract Concentration (ppm)
TPH (diesel fuel)	30–600	300
Anthracene	1.125–22.5	11.25
BA	1.5–15	7.5
BaP	0.9–9	4.5

2.3.1.6 Plant Biomass Analysis

Prior to the destruction of the plant, the physical parameters of both the root and shoot were measured and recorded. Once separated from the soil, the plants were kept in cold storage until analysis was performed. The plants were washed in tap water to remove any soil adhered to the roots and blotted dry with paper towels. The roots were then separated from the shoots, and each was weighed separately for wet biomass. The root surface area for each vegetated column was determined by using a Delta-T SCAN image analysis system. The system utilized a flat-bed digital scanner, a PC-compatible computer, and image integration software. The root sample was placed on a clear transparency sheet and spread out to minimize overlap of the root structure. The transparency sheet was then placed on the scanner panel and covered with a white background. The image was then scanned and recorded as a TIFF digital bitmap image. The image was edited using image-editing software to remove any unwanted marks picked up from the transparency sheet. The Delta-T image software then processed the image by counting the number of dark pixels that represented the roots. Using the size of a pixel, and the number of pixels in the image, the program calculated the total surface area shown in the two-dimensional image. To convert the two-dimensional image area to a three-dimensional surface area, the measured area was multiplied by p (3.14).

2.3.1.7 Microbial Enumeration

Microbial assessment consisted of viable microbial counts using spread plates at every takedown. Spread plates were made from tryptic soy agar (TSA). Ten grams of soil at room temperature was added to 90 ml of sterile 0.2% tetrasodium pyrophosphate ($Na_4P_2O_7$) in a sterile 160-ml serum bottle. The bottle was capped with a rubber stopper and placed on a shaking table for 30 min. A 1-ml volume of the soil solution was serially diluted. A 0.1-ml volume of the 10^{-4} and 10^{-5} dilutions was spread evenly upon the TSA plates in triplicate. The completed plates were stored at room temperature for 60 hr. The number of colonies was counted at 24, 48, and 60 hr. The plate dilution that resulted in consistent results between 30 and 300 colonies per plate was chosen for enumeration.

2.3.1.8 Quality Assurance and Quality Control

All chemical extractions were done with two blank samples per analysis of 12 columns. Samples were duplicated at a rate of one duplicate per nine samples (11.11% duplication). Two chemical compounds were added to the extracts. Tetracosane, applied at 5 ppm, was used to measure extraction efficiency as a matrix spike. A 5-ml volume of 1000-ppm androstane was injected into each GC vial as an internal standard.

2.3.1.9 Statistical Analysis

Raw data from root surface-area measurement, GC results of soil extractions, and micro-bial enumeration were analyzed for significant differences. The analysis of variance (ANOVA) method was used to determine the LSD value at a p value of 0.05 or 5%. The LSD was then used as a measure of statistical significance for the data sets. Statistics for all other sets of raw data were limited to measuring means and standard deviations.

2.3.2 Results

2.3.2.1 Plant Biomass Analysis

Columns were taken down and analyzed at 7, 14, and 21 weeks. Soil samples were also taken at week zero. At each takedown, plants were separated into root and shoot biomass and weighed. Figures 2.7 and 2.8 show the shoot and root biomass, respectively. Each pot began with four plants; at week 21, either three or four plants survived in all pots. At each soil depth, the entire mass of plant material was taken as the sample. Both shoot and root biomass were greater for plants in the uncontaminated soil than for plants grown in contaminated soils. Table 2.11 shows the ratios of uncontaminated to contaminated biomass weights for shoots and roots at all three soil levels at 21 weeks.

After weighing, the root samples were scanned for surface area. The plants grown in uncontaminated soils consistently had higher root surface area than plants grown in the contaminated soil for a given soil layer (Figure 2.9). In the uncontaminated soil, the surface area of the top, middle, and bottom root samples decreased with depth at each sampling date. In the contaminated soil, the root surface areas in the top and middle soil layers were nearly equal at 21 weeks, while the root surface area in the bottom layer was approximately 21% smaller.

Using a three-way ANOVA ($p = 0.05$), the LSD between the measured surface areas was determined to be 19,400 mm². Statistically significant effects included all three main effects (contamination, soil depth, and time) and the interaction between contamination and time. The behavior of root and shoot mass, as well as root surface area, showed that the presence of contaminants in the soil impaired the normal growth of vegetation. The drop in shoot

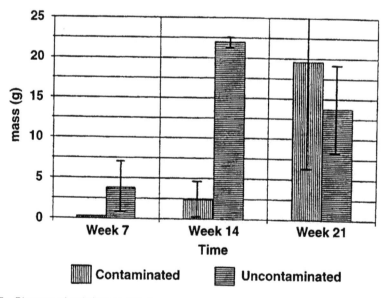

Figure 2.7   Biomass of switchgrass shoots.

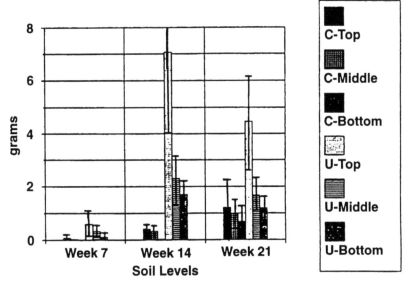

Figure 2.8   Biomass of switchgrass roots at three soil depths: contaminated (C) vs. uncontaminated (U).

biomass for the uncontaminated soils at week 21 was most likely due to the onset of senescence and the effects of predation by insects. The fact that plants grown in contaminated soils did not exhibit this decrease in shoot biomass suggested that the plants grown in contaminated soil may experience a retardation in their life cycle progression. This observation was supported when plants grown in uncontaminated soil produced seed nearly 2 weeks before those plants grown under contaminated conditions.

Plant material sampled at 21 weeks was analyzed for anthracene, BA, and BaP. GC of extracts from both the roots and shoots revealed no discernible peaks for the target compounds. Peaks appearing at other times throughout the analysis appeared randomly in samples grown in uncontaminated as well as contaminated soil. Uptake of contaminants was either below detection limits or the contaminant was metabolized to other compounds within the plant.

2.3.2.2 Soil Contaminant Analysis

The concentration of TPH in the soil varied between 1144 and 1485 mg/kg at week 0 (Figure 2.10). It can be seen that at week 0, the concentration of TPH was far below the target (added) level of 2000 mg/kg. This may be due in part to the volatilization of some components of the diesel fuel, but is probably due to inefficient extraction. The variability of the

Table 2.11 Biomass Ratios for Plant Samples at 21 Weeks	
Sample Type	Ratio (g UC/g C biomass)[a]
Shoot	1.42
Roots	
Top layer	17.24
Middle layer	7.19
Bottom layer	85.0

[a] UC = uncontaminated, C = contaminated.

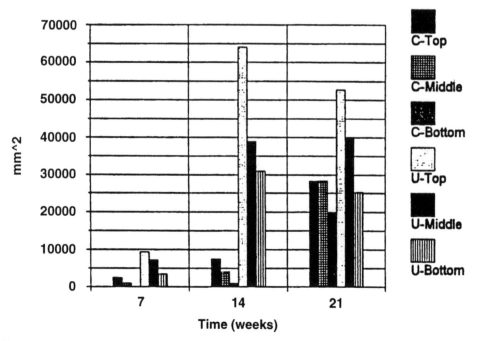

Figure 2.9   Root surface area at three soil depths: contaminated (C) vs. uncontaminated (U). LSD = 19,419.59.

initial concentrations at each soil layer was relatively high. At week 7, the concentrations in each soil layer for either treatment (vegetated or unvegetated) showed significant differences. The top layers had significantly lower TPH levels than those in the middle layer, while the levels in the middle layers were significantly lower than those in the bottom layer. At 21

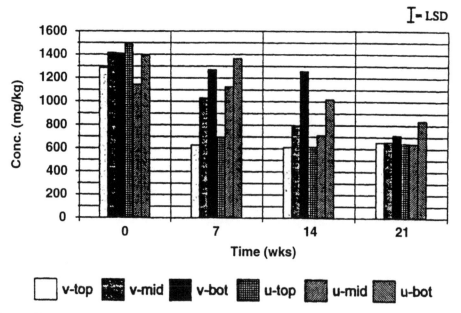

Figure 2.10   TPH concentration in the soil vs. time.

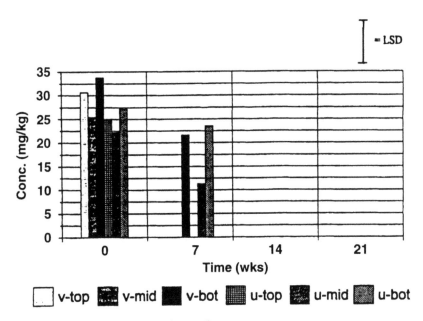

Figure 2.11   Soil anthracene concentration vs. time.

weeks, there were no significant differences in the concentrations between the two treatments for the top, middle, or bottom layers.

An ANOVA of the soil extraction data ($p = 0.05$) resulted in an LSD of 138 mg/kg or, when analyzed as a percent relative to week 0, LSD = 10.13% at $p = 0.05$. Significant effects were due to the design variables (factors) of soil depth and time. Significant interactions between different factors included all possible two- and three-factor combinations of the three main factors (depth, time, and vegetative treatment). The difference between TPH concentration in the vegetated bottom depth and concentration in the unvegetated bottom depth was slightly significant ($p = 0.10$). Therefore, under less discriminating conditions, there is a positive effect of vegetation on the degradation of TPH compounds in soil.

At week 0, the concentration of anthracene in the soil was between 22 and 33 mg/kg (Figure 2.11), which was far below the target level of 75 mg/kg. There was a great deal of variability in the concentrations between soil depths and vegetative (V) treatments. At week 7, a significant drop in concentration was observed. In the vegetated columns, anthracene could not be detected in the top and middle soil depths. In the unvegetated (U) columns, the top layer had no measurable anthracene, while the compound was still detectable in the middle and bottom layers. At weeks 14 and 21, no anthracene was detected in any of the soil depths for either treatment.

ANOVA analysis of anthracene concentrations in the soil provided LSD values of 9.1 mg/kg and 34.7% ($p = 0.05$). The high rate of degradation was an important factor in the large variability of the data. Significant factors included the effects of time and soil depth. Significant interactions were those between soil depth and time.

The concentration of BA in the soil at week 0 ranged from 33.8 to 35.9 mg/kg (Figure 2.12). The target value of BA was calculated to be 50 mg/kg, while the measured concentrations measured at time 0 were closer to 35 mg/kg. These values exhibited a low variability between treatments and soil depths.

At week 7, BA decreased significantly in the top and middle soil depths of the vegetative treatments, but the decrease in the bottom layer was not statistically significant. At week 14, the trend continued. At week 21, the BA concentration in the vegetated bottom soil layers

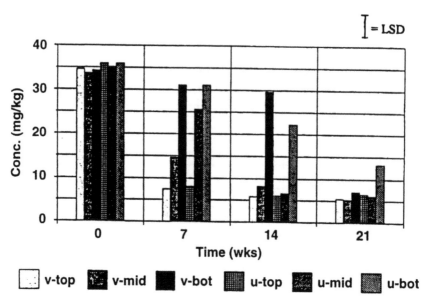

Figure 2.12   BA concentration vs. time.

declined significantly and reached levels not statistically different from the top and middle layers. The BA concentration decreased in the unvegetated bottom layer, but it was still significantly higher than the concentrations in the middle and top layers. Vegetation apparently had a significant effect on the degradation of BA in the bottom soil layer, but not in the middle and top layers.

ANOVA of the BA contamination data yielded the LSD value of 5.39 mg/kg and 15.44% (p = 0.05). ANOVA also indicated that the factors depth and time produced significant effects, while significant interactions were observed between depth and time, vegetative treatments and time, as well as the combination of all three factors (time, depth, vegetative treatment).

The soil extracts contained BaP at concentrations of 19.08 to 19.76 mg/kg at week 0 (Figure 2.13), which was less than the target level of 30 mg/kg. There was very little variability in the measured concentrations of BaP at week 0. The behavior at weeks 7 and 14 was similar to that of BA and TPH. The top soil layer possessed the lowest levels of BaP, while the middle and bottom layers possessed the second lowest and highest levels, respectively. Statistical analysis indicated that at week 7, the only significant differences were between the top and middle layers and between the top and bottom layers for both the vegetated and unvegetated columns. At week 14, significant differences existed only in the case of the bottom soil layers relative to the top and middle layers. These differences were significant for both vegetative treatments. The results from week 21 showed a significant difference in BaP concentration only in the bottom soil layer of the unvegetated column. The BaP concentration of the bottom layer in the vegetated treatment at 21 weeks was not significantly different from the middle and top layers of the same column. The vegetative treatment had a significant effect on the degradation of BaP, but only near the bottom of the rooting zone within the column. Observable differences in BaP levels between the vegetated and unvegetated columns were seen for all soil levels, but they were not all significant.

The ANOVA performed for BaP concentrations yielded LSD values of 2.82 and 14.35 measured as concentration and percentage, respectively (p = 0.05). Significant effects were

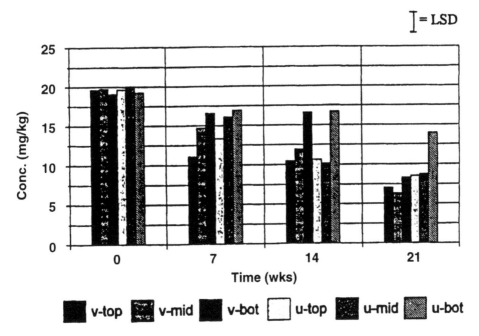

Figure 2.13   BaP concentration vs. time.

found to be soil depth, vegetative treatment, and time. Significant factor interactions were between depth and time, as well as vegetation and time.

Generally, the petroleum contaminants degraded more quickly near the top of the columns. The effect of vegetation on the contaminants was most noticeable near the bottom of the columns, with the heavier PAH compounds responding better to vegetation than the compounds measured as TPH components.

2.3.2.3 Analysis of Root Surface Area vs. Contamination

To evaluate potential relationships between root surface area and the degradation of soil contaminants, several trials of regression analysis and data transformation were attempted. The two methods of analysis that were used were linear regression analysis and multiple linear regression analysis. Data transformations of the independent variable (surface area) included using natural logarithms, inverse values, and the square, cube, and other root values of the surface area. Transformations of the dependent variable included using percent degradation of initial value of concentration of contaminant. All these transformations were compared along with the untransformed values of root surface area (square millimeters) and contaminant concentration (milligrams per kilogram) through linear regression and multiple linear regression analysis.

Through iteration of regression and multiple regression results, the best linear fit was made using a direct linear regression between percent degradation as the dependent variable and the root surface area raised to some power X as the independent variable. The expression is shown below:

$$D = (A^x) + b$$

where D = percent degradation of contaminant (%), A = surface area of root system (mm^2), x = empirical constant (for a specific compound), and b = y-intercept (constant).

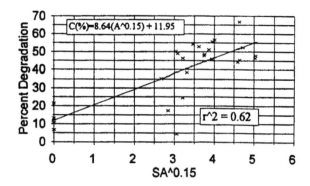

Figure 2.14   Regression of root surface area vs. TPH percent degradation: actual (x) vs. best-fit (—) data.

The value of X varied with each specific contaminant, and R^2 values of these linear fits ranged from 0.62 to 0.68, increasing with the recalcitrant character of the specific compound. Figures 2.14 to 2.16 show the shape of the best-fit regression line along with the form of the equation for each specific contaminant. Table 2.12 provides the value of the empirical constant X for specific compounds. Anthracene was not included in this analysis due to the low number of data points available. Multiple regression analysis provided similar results, but no significant improvements were observed in the R^2 value of the regression line.

A better fit of the data may have occurred if the experiment had continued longer than 21 weeks. Figures 2.14 to 2.16 show that as surface area increased, the fit of observed data to the regression line improved. Better results also may have been obtained if a greater number of plants was grown in the columns to reduce the variability.

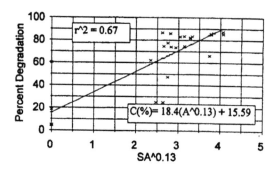

Figure 2.15   Regression of root surface area vs. BA concentration: actual (x) vs. best-fit (—) data.

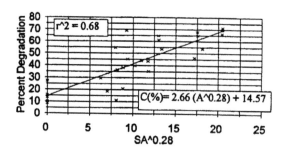

Figure 2.16   Regression of root surface area vs. BaP concentration: actual (x) vs. best-fit (—) data.

Table 2.12 Values of Empirical Constants and R^2 Values for Root Surface vs. Compound Degradation

Compound	Constant	R^2
TPH	0.15	0.62
BA	0.13	0.67
BaP	0.23	0.68

2.3.2.4 Microbial Enumeration

The enumeration for time 0 soil was averaged over four subsamples taken prior to wetting, contaminating, or adding the nutrients to the soil. It was assumed that the population distribution in the soil at time 0 was homogeneous and that approximately 436,000 colony-forming units per gram of dry soil were found in each of the columns originally.

The unvegetated uncontaminated (UU) soil had a fairly homogeneous distribution of microbes at week 7 (Figure 2.17). The vegetated uncontaminated (VU) column had significantly more microbes in the top layer than in the middle and more in the middle than in the bottom layer. The unvegetated contaminated (UC) column had a larger number of microbes in the bottom layer than in the middle and more in the middle than in the top. Finally, the vegetated contaminated (VC) soil possessed a fairly homogenous distribution of microbes, similar in nature to the UU treatment, but with approximately twice the numerical value.

The relationship between microbial numbers and soil depths for the VU columns and the UC columns at week 14 was the same as observed at 7 weeks (Figure 2.18). The VC column changed significantly from week 7. At week 14, the numbers in the upper layer declined significantly from week 7, leading to a distribution of microbial numbers that resembled the UC column. The bottom layer possessed much higher microbial numbers than did the middle, which possessed more than the top layer.

The microbial distribution within the UU columns at 21 weeks was relatively homogeneous (Figure 2.19). The VU columns still possessed higher populations near the top, but by a barely significant margin. The UC column exhibited the same trends as were seen at week 14. The VC column also continued with the trends exhibited at week 14. The bottom layers had more microbial numbers than the middle layer, and the middle had higher numbers than the top. However, there were significant declines in microbial numbers in all columns and depths except the UU by 21 weeks.

Statistical analysis of the data was performed using a four-way ANOVA procedure. The LSD of the microbial numbers was calculated to be 1.41×10^6. Statistically significant effects

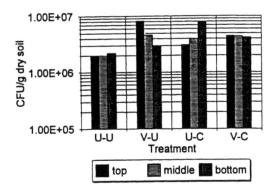

Figure 2.17 Microbial populations at three soil depths at week 7. LSD = 1.41E+06.

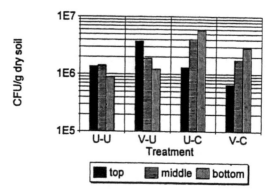

Figure 2.18   Microbial population at three soil depths at week 14. LSD = 1.41E+06.

included the effect of time as a main effect, as well as the interactions of depth and contamination, vegetation and contamination, depth and vegetation, and finally the interaction between all four main factors (depth, vegetation, contamination, and time).

The rhizosphere effect may be influenced by the presence of contaminants in the soil. Contamination appeared to alter the distribution of the microbial biomass within the column. The data indicated that the magnitude of the plant effect on microbial activity was less than the influence of contamination and soil moisture conditions. These factors may have become the dominant forces in determining the distribution of soil microbes, perhaps lessening the potential impact of phytoremediation near the soil surface. The effect of soil moisture cannot be overlooked when considering these relationships. It has been well established that soil moisture content has a strong effect on microbial activity, maximizing activity at around 60% water-filled pore space (Paul and Clark, 1989).

Soil moisture content in the VU columns suggested that the plants created a gradient that was opposite to the gradient seen in the UU column and the UC column (Table 2.13). The effect of vegetation on water content was not as significant in the VC columns as in the VU columns. The UU columns had very moderate increases in soil moisture as depth increased, while the UC columns had a much stronger gradient.

Vegetation appeared to cause an upward water gradient. In the unvegetated columns, water accumulated in the lower levels (an effect of gravity). Vegetative effects on the soil structure should also be considered. The creation of macropores through root growth may

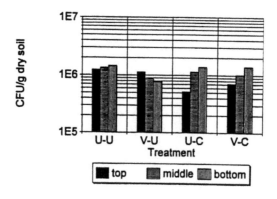

Figure 2.19   Microbial population at three soil depths at week 21. LSD = 1.41E+06.

Table 2.13 Average Soil Moisture at Various Soil Depths

Soil Depth	VU Column	VC Column	UU Column	UC Column
Top	18.8%	15.7%	13.2%	14.0%
Middle	14.3%	15.4%	13.9%	16.1%
Bottom	14.9%	15.6%	14.5%	18.7%

allow for more rapid transport of oxygen, water, and nutrients to the lower soil layers. These physical effects may also play an important role in stimulating the degradation of organics at the lower soil depths.

2.3.2.5 Leachate Analysis

Immediately after the soil columns were filled, the soil was saturated with water and allowed to settle. Any leachate that was produced (week 0) was collected. After extraction of the hydrocarbons from the water, GC analysis was performed to determine the concentrations of contaminants in the leachate. TPH was measured first. The total loss of contaminant from the columns ranged from 0 to 0.015% of the total initial mass of contaminant in the soil. Considering the solubilities of some of the main components of diesel fuel (0.07 to 3 mg/l) relative to anthracene (0.03 mg/l), BA (0.014 mg/l), or BaP (0.003 mg/l), the losses of these compounds were insignificant relative to the concentrations in the soil (Heath et al., 1993). No more leachate was produced for the duration of the experiment.

2.3.3 Conclusions

2.3.3.1 Root Surface Area

The comparisons of root surface area to the degradation of organic contaminants resulted in a linear-fit model utilizing an empirical constant that was unique for each of the three contaminants compared in the analysis. Using the R^2 statistic as a measure of fit, the best fit provided R^2 values between 0.62 and 0.68 depending on the contaminant investigated. The results may be largely a function of the plant species used. For plants other than switchgrass, the relationships investigated may be significantly different, including the values of the empirical constant and the statistical fit of the regression line.

The fit of the regression line appeared to improve with increasing molecular weight of the PAH and with increasing recalcitrant nature of the organic compound. The more recalcitrant, heavier compounds with more complex ring structures are the least responsive to conventional bioremediation and are therefore targets of phytoremediation. For compounds that are highly responsive to bioremediation without vegetation, a poorer relationship between root surface area and degradation is expected.

These results establish a beginning for further investigations of the relationship between plant parameters and the effectiveness of phytoremediation in degrading organic contaminants. Further investigation is necessary if these relationships are to be utilized in computer models and design of remediation schemes.

2.3.3.2 Microbial Enumeration

The presence of contaminants had a strong effect on the distribution of bacteria and other microorganisms. After 7 weeks, the distribution of microbial numbers in both the UU and VC columns was homogeneous, but the VC columns had significantly higher numbers in all layers, probably as a result of the carbon and energy sources provided by the contaminants and the plants. Soil contaminants quickly disappeared from the upper levels of the unvegetated and vegetated columns and remained high in the lower soil layers; the lower layers may have been oxygen limited. After 14 weeks, microbial numbers were lower in the upper levels of both the vegetated and unvegetated columns where contaminant concentration had declined dramatically than in the lower levels where contaminant remained. The presence of vegetation in uncontaminated soil helped to redistribute more water to the top and middle layers of the soil columns, thereby redistributing the microorganisms, which are sensitive to soil moisture. In unvegetated soils, soil moisture tended to accumulate near the bottom layers of the soil column. The relationship among contamination, microbial numbers, and soil moisture may be very important in assessing the viability of an *in situ* biological remediation strategy.

2.3.3.3 Degradation of TPH and PAHs

Soil contamination decreased over time in both the vegetated and unvegetated columns. In all cases, the degradation was highest in the top soil layers, while the bottom layer showed the least amount of degradation over the 21-week period. Diesel fuel (measured as TPH) in the top and middle soil layers degraded very quickly, while the bottom layer possessed significantly higher TPH levels at every analysis time. At 21 weeks, the bottom layer of the vegetated soil column did not possess significantly less TPH than the bottom layer of the unvegetated column.

Anthracene degraded very quickly from all the columns in the experiment. No anthracene was measured in any column at week 14. At week 7, anthracene was detected in both the middle and bottom soil layers of the unvegetated columns, while the vegetated columns possessed anthracene only in the bottom soil layers.

BA and BaP showed significantly higher degradation in the bottom layers of the vegetated columns after 21 weeks than in the bottom layers of the unvegetated columns. The contamination levels in the top and middle layers were not significantly different at week 21. Again, the top and middle layers exhibited faster degradation in both column types than did the bottom soil layers.

Results from the soil extractions indicate that vegetation has a more significant impact on PAH degradation in deeper soils than in the top layers. The impact of vegetation at that depth may be stronger due the introduction of root exudates into the soil, the creation of macropore flow into lower soil layers, and enhanced oxygen diffusion into the lower soil depths. The relationship among soil contaminants, microbial activity, and soil moisture may also play a role here. Most likely, it is a combination of these factors that causes the response of organic contaminants to vegetation.

2.4 GERMINATION ASSESSMENT FOR CRANEY ISLAND SOIL

Prior to planting the Craney Island field study site, plant species that would have a high probability of successful establishment were sought. Several factors enter into species selection, including adaptation to the climate and soils of southeastern Virginia, tolerance to or minimal growth reduction in the target contaminated soil, shallow rooting system compatible with the soil depth, and design of the bioremediation unit.

Although a replicated greenhouse trial testing the germination and growth characteristics of candidate species would have been desirable, limited access to contaminated soil samples provided only enough soil for a small study of seven candidate species. The objective of this study was to determine if Bermuda grass, tall fescue, and white clover would establish and begin rooting in the Craney Island contaminated soil.

2.4.1 Materials and Methods

Seven vegetation treatments were grown in three soils in unreplicated 10-cm plastic pots. The experiment was conducted in a growth chamber set on a 14-hr day length with a 26°C day temperature and 16°C night temperature. The vegetation treatments listed below included two seeded pots and five pots established from sod plugs obtained from the Kansas State University turfgrass research program:

1. 100 seeds — K-31 tall fescue
2. 100 seeds — Dutch white clover
3. Tall fescue sod — variety Marksman
4. Tall fescue sod — variety K-31
5. Bermuda grass sod — variety Midiron
6. Bermuda grass sod — variety Midlawn
7. White clover/bluegrass sod — variety unknown

One pot of each treatment was established in three soils: uncontaminated Eudora silt loam obtained locally, contaminated soil sample CIT from Craney Island, and contaminated soil sample CIP from Craney Island. The CIT soil had a moderate amount of sand mixed with clay. The CIP sample was a heavy clay with a strong odor suggesting high levels of contaminants. It was unclear at the time of this experiment which soil sample would be representative of the field study.

The experiment was run from June 29 to July 26, 1995. Pots were watered twice weekly and fertilized once weekly with Miracle Gro™. At the conclusion of the study, plant number was counted in the seeded pots. Soil was washed from the roots to examine root growth. For seeded pots, roots and shoots were dried and weighed. Only roots penetrating contaminated soil were dried and weighed for pots established from sod.

2.4.2 Results

In general, plants germinated and grew in the contaminated soils. There was no strong evidence of serious phytotoxic effects of the contaminated soils. Only seeded white clover showed apparent stunting of growth in contaminated soil. White clover established from sod showed the best rooting into contaminated soil of all treatments. Clover roots established from sod were nodulated, suggesting the potential for nitrogen fixation. Clover in the CIP soil sample had the best rooting of all treatments in the study. Tall fescue appeared to grow well in all soils with only a small reduction in plant height in the CIP soil sample. Tall fescue sod was beginning to show root growth into contaminated soil. Bermuda grass foliage appeared to develop well; however, the Bermuda grass showed the poorest rooting in all soils. The temperatures may have been too cool for healthy Bermuda grass establishment. Table 2.14 presents all the data on root production and plant number.

The results suggested that the chosen treatments, Bermuda grass sod, tall fescue seed, and white clover seed, would be able to grow in contaminated soil. Sod may be a useful way to establish vegetation rapidly using mature plants rather than seedlings. However, it is important to monitor rooting from the sod into the contaminated soil during the field study period.

Table 2.14   Root and Shoot Weight for Tall Fescue, Clover, and Bermuda Grass Grown in
Contaminated and Uncontaminated Soil During Initial Greenhouse Study

Vegetation	Stock	Soil	Root wt (g)	Shoot wt (g)	Plant no.
Tall fescue	Seed	Uncontaminated	0.208	0.350	67
(K-31)		CIT	0.235	0.209	85
		CIP	0.145	0.267	84
White clover	Seed	Uncontaminated	0.023	0.078	25
(Dutch white)		CIT	0.039	0.093	56
		CIP	0.023	0.010	15
Tall fescue	Sod	Uncontaminated	0		
(K-31)		CIT	0		
		CIP	0.061		
Tall fescue	Sod	Uncontaminated	0.033		
(Marksman)		CIT	0.028		
		CIP	0.031		
Bermuda grass	Sod	Uncontaminated	0		
(Midiron)		CIT	0.025		
		CIP	0		
Bermuda grass	Sod	Uncontaminated	0		
(Midlawn)		CIT	0.024		
		CIP	0		
White clover/	Sod	Uncontaminated	0.014		
bluegrass		CIT	0.028		
		CIP	0.056		

CHAPTER 3

Field Study

3.1 SITE DESCRIPTION

The Craney Island Fuel Terminal (CIFT) in Portsmouth, VA is the Navy's largest fuel storage facility in the U.S. The CIFT consists of over 1100 acres (445 ha) of both underground and aboveground fuel storage tanks. The phytoremediation study was conducted in the Atlantic Division, Naval Facilities Engineering Command biological treatment cell located at the CIFT. The bioremediation treatment cell is approximately 15 acres (6.1 ha) and is underlain by layers consisting of a compacted clay base, a synthetic geogrid (for stabilization), sand layers, and a polyethylene liner. The cell is bermed on all sides and uses sumps and pumps to collect and remove irrigation and storm water.

The contaminated material used in this study was excavated in 1995. Lagoons approximately 200 ft (61 m) × 70 ft (21.4 m) had been used at the CIFT from 1940 to 1978 for gravity oil–water separation of ballast and bilge waste from ships. The petroleum-hydrocarbon-contaminated lagoons were filled with soil in 1980. In 1995, a remediation plan was developed, and the contaminated sediments were removed for treatment in the CIFT biotreatment cell. Petroleum-contaminated soil is treated by bioremediation in the biotreatment cell by first placing 18 in. (45.4 cm) of soil in the cell and mixing using construction equipment. After initial mixing, aeration, irrigation, and fertilization are performed to encourage bioremediation of the contaminants.

The area where the phytoremediation study took place is a portion of the bioremediation treatment cell. A 0.5-acre (0.20-ha) area was marked off and filled with contaminated soil to a depth of 2 ft (0.61 m) (Figure 3.1). After intensive sampling and characterization, plants were established on the study area. Target compounds were measured over time to evaluate the efficiency of phytoremediation. The contaminated soil adjacent to the phytoremediation study area was treated using bioremediation and removed from the biotreatment cell in the fall of 1996.

3.2 INITIAL SITE ASSESSMENT

Before establishment of vegetation, soils were intensively sampled in 12-in. (30.5-cm) increments to the underlying sand layer (approximately 2 ft [0.61 m]). Soil samples were analyzed for initial concentrations of contaminants and characterized using standard soil tests.

3.2.1 Soil Characterization

An important aspect of soil remediation research is the routine determination of important chemical and physical properties that affect water, plants, and microorganisms. In September

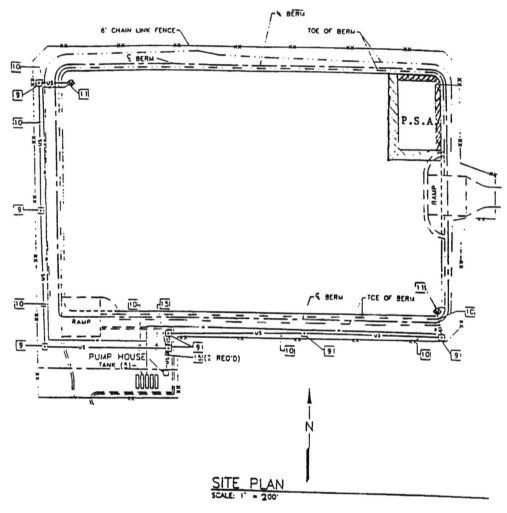

Figure 3.1   Craney Island biological treatment facility. P.S.A. = phytoremediation study area. Approximately 180 ft × 100 ft. Exposed buffer zone = 20 ft (□). Edge buffer zone = 10 feet (☒).

1995, a thorough characterization was made, and only those properties that could change over the short term were determined subsequently (Tables 3.1 to 3.3).

The texture (sandy loam) was indicative of relatively low water-holding capacity, but the high organic matter content helped counterbalance the coarse texture. The electrical conductivity (EC) indicated slightly elevated salt content, but not enough to present plant growth problems. Similarly, the sodium adsorption ratio (SAR) was slightly elevated compared to soils that are unaffected by sodium, but the SAR was not high enough to indicate negative effects. The low cation exchange capacity (CEC) reflected the low clay content, and the moderate Bray-P suggested the need for phosphorus fertilization to maintain maximum productivity of the vegetation.

In 1996 and 1997, only the pH and residual inorganic N and P were measured. The pH was stable, as one would expect for this type of project. The Bray-P remained in the moderate range despite the regular fertilization schedule, indicating that the site was not fertilized excessively. Residual N was always low to moderate.

Table 3.1 Initial Soil Characterization
Results, September 1995

Parameter	Value
pH	7.4
EC	4.0 mmhos/cm
NO_3–N	<0.1 mg/kg
NH_4–N	2.70 mg/kg
Bray-P	20.5 mg/kg
CEC	7.4 meq/100 g
Organic matter	4.4%
Ca	22.5 meq/l
Mg	14.8 meq/l
Na	12.0 meq/l
K	2.5 meq/l
SAR	2.8
Sand	60%
Silt	21%
Clay	19%
Texture	Sandy loam
Total organic carbon	1.8%
Solids	78.2%
Salt rank	Low

Table 3.2 Soil Characteristics of the Different Plot Treatments in the
Phytoremediation Study Area, August 1996

Plot	pH	Bray-P (mg/kg)	NH_4–N (mg/kg)	NO_3–N (mg/kg)
White clover	6.2	23	14.4	8.5
Tall fescue	6.9	13	5.0	0.7
Bermuda/rye	6.8	13	3.5	0.7
Unvegetated	5.9	13	13.5	8.5

Table 3.3 Soil Characteristics of the Different Plot Treatments in the
Phytoremediation Study Area, October 1997

Plot	pH	Bray-P (mg/kg)	NH_4–N (mg/kg)	NO_3–N (mg/kg)
White clover	6.5	27	5.3	5.7
Tall fescue	6.3	24	3.9	4.9
Bermuda/rye	6.5	21	2.5	0.3
Unvegetated	6.5	22	2.6	0.3

3.2.2 Initial Contaminant Analysis

The initial samples were taken and analyzed for total petroleum hydrocarbons (TPHs) using shaking extraction with quantification by infrared spectroscopy. Preliminary contaminant analysis of TPH in the phytoremediation area showed that the degree of spatial variability was smaller than that expected in natural soils (Tables 3.4 and 3.5). The mean value was 4551 mg/kg with a standard deviation of 1045 mg/kg. The highest and lowest values were 7923 and 3036 mg/kg, respectively. A log transformation of the TPH data suggested that the univariate description could be well approximated by the normal distribution. Preliminary geostatistical analysis indicated that the range of the omnidirectional variogram was between 35 and 40 ft.

Surface and contour plots of the TPH data showed that there was no distinct trend from one side of the field to the other (Figures 3.2 and 3.3). There were two high-concentration

Table 3.4 Initial Sampling Results for TPHs in Phytoremediation Study Area

North[a]	5.0	20.0	27.5	35.0	36.5	44.0	44.5	48.0	50.0
West[b]									
7.5	5383[c]	4859		4284					3037
22.5	5690	4337		3409					2855
37.5	4127	3247		3915					3000
52.5	4605	4653		4434					3683
67.5	4051	4746		5060					4121
82.5	3687	4356		4730					4757
90.0			5233				5005		4867
92.5								3986	
97.5	4036	4092		4405		4148			
99.0									4273
105.0						3628			
108.0									4530
112.5	3610	4648		4433					4753
127.5	4381	5741		4897					4849
142.5	4315	4666		6257					7857
157.5	4169	4519		4194					5035
172.5	3948	3821		4401					7531
North	51.5	55.0	56.0	57.0	63.0	65.0	72.5	80.0	95.0
West									
7.5						3499		3680	3167
22.5						3497		4762	3836
37.5						5054		3621	4416
52.5						4147		4686	7924
67.5						4496		3758	3570
82.5						3069		4743	3865
91.0		5079							
94.0				4725					
97.5	4631				3109	3319		3330	3756
105.0			3933				3473		
112.5						4103		4861	4071
127.5						3405		4310	6492
142.5						3915		6290	8981
157.5						6800		6996	6329
172.5						5991		7154	5580

[a] Distance from northern border in feet.
[b] Distance from western border in feet.
[c] Units are milligrams per kilogram.

regions, with the major one being in the northwest corner of the field. The other distinct peak was also on the north side of the field. Otherwise, these figures indicated that high and low values were fairly well mixed. Since the TPH values stayed within the same order of magnitude, a complete randomized block design was adopted.

Table 3.5 Time 0 TPH Concentration as Determined by Environmental Testing Services, Norfolk, VA

	mg/kg			
	Unvegetated	Bermuda/Rye	White Clover	Tall Fescue
Replicate plot 1	1140	1380	723	885
Replicate plot 2	6620	882	1030	992
Replicate plot 3	1230	1270	1290	1270
Replicate plot 4	1100	1070	509	463
Replicate plot 5	697	534	1160	556
Replicate plot 6	575	860	484	1320

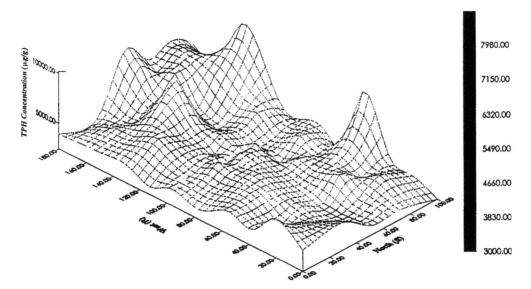

Figure 3.2 Surface plot of TPH concentrations (µg/g).

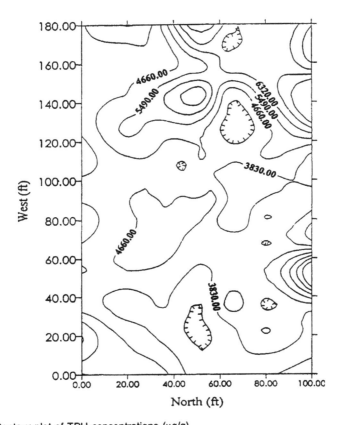

Figure 3.3 Contour plot of TPH concentrations (µg/g).

52

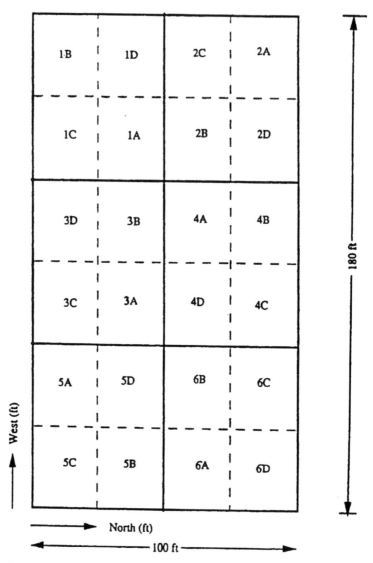

Figure 3.4 Schematic layout of randomized block diagram. The number stands for the block number, and the letter stands for the treatment.

3.3 STUDY AREA DESIGN

Based on the size of the field, the number of treatments (four in this case), the nature of spatial variability, and other considerations having to do with management of each plot, six blocks (or replicates) were used. The placement of these blocks and the treatments within each block are shown in Figure 3.4. The treatments were designated as A (white clover), B (unvegetated control), C (tall fescue), and D (Bermuda/rye grass). There were two major reasons for adopting this design in the present study. The choice of six blocks with four treatments restricted the size of each treatment plot to about 30 ft (9.14 m) × 25 ft (7.62 m) (ignoring the borders between the treatment plots), which was less than the spatial correlation lengths of 40 ft (12.2 m). It was considered important to keep each block as a contiguous unit (and not fragmented), so that each block would encompass more than a correlation length in each direction. The second criterion was to minimize the variation within each block to the extent possible. The relatively high values in the northwest region necessitated a complete block.

Four samples were collected and analyzed from each plot. The four sampling points within the blocks were placed to reduce any edge effects and to minimize any influence from neighboring plots. These four values were averaged to represent the concentration within each treatment plot. The differences between the average values of the treatments between different time periods were treated as a measure of performance, and an analysis of variance was performed to compare the performance of each treatment with respect to the control.

3.3.1 Sampling Plan

Throughout the study period, soils were sampled. A 1-in. (2.54-cm) soil probe was used for sampling to depths up to 24 in. (61 cm). Soil samples were placed in sterile containers, labeled according to sample location and depth, and transported to Kansas State University (KSU) in coolers for analysis. The samples were analyzed for target contaminants at the end of the study. Aboveground plant biomass was assessed, and plant tissue was tested for contaminants. Also, soil solution samplers (vacuum pore-water samplers) were installed in several phytoremediation plots to assess leaching of contaminants. These samplers collected soil solution at various depths between the soil surface and the top of the underlying sand layer.

3.4 ANALYSES OF SAMPLES

Soil, water, and biomass samples were collected and analyzed. Details of the analyses performed on the samples are provided below.

3.4.1 Soil Contaminant Analysis

As described in the laboratory results section (Section 2.1), the combination of soil:solution ratio, solvent, and number of sequential shaking extractions that removed the same amount of TPH as soxhlet extraction of 5 g of soil with 100 ml of dichloromethane (also see Appendix 1) was determined systematically. In the current extraction procedure, 2 g of soil was air dried, ground to pass a 40-mesh sieve, shaken with three sequences of 10 ml of dichloromethane, and then analyzed for TPH by gas chromatography (GC) with flame ionization detection (FID). The shaking method was used on a routine basis, and soxhlet was used only for samples from the initial (October 1995), midproject (August 1996), and final sampling (September 1997) times.

3.4.2 Analysis of Target Compounds

Many physical, chemical, and biological changes occur in petroleum products when they are released into the environment. Low-molecular-weight alkanes will be depleted by volatilization; higher molecular weight, straight-chain alkanes can be depleted by bacterial degradation (Douglas et al., 1992); and some compounds, such as the polycyclic aromatic hydrocarbons (PAHs) with three or more rings, are strongly adsorbed by soil (Knox et al., 1993) and are resistant to weathering. Some of the strongly adsorbed PAHs are also potentially toxic or carcinogenic, and the quantification of these compounds is often used to assess the potential risk to humans and animals associated with petroleum hydrocarbon contamination of soil.

Another important group of chemicals associated with petroleum is referred to as "biomarkers" because they are naturally occurring yet are highly resistant to degradation. Originally used in the petroleum industry to trace oils to their sources and to understand their genesis, Douglas et al. (1994) and Prince et al. (1994) used $C_{30}17(H)21(h)$-hopanes as internal standards to normalize the loss of petroleum contamination through biodegradation. It should be noted that biomarkers degrade at least to some extent, but these changes are predictable and quantifiable (Kaplan et al., 1996). Because biomarkers are found only in oil, their concentrations in contaminated soil can be compared to the concentration of TPHs and other specific compounds to evaluate biodegradation, account for spatial variability of field data, and increase the accuracy of the assessment of remediation. The objective of this portion of this study was to monitor the TPH, PAH, and biomarker (hopane and norhopane) concentration changes and biodegradation in the Craney Island soils.

3.4.2.1 Soil Extraction and Column Chromatography Methods

Ten grams of soil was extracted by soxhlet for 3.5 hr using 100 ml dichloromethane. Ten grams of silica gel (100 to 200 mesh, activated at 110°C for 24 hr) was packed into a chromatographic column and washed with two successive 20-ml aliquots of hexane. Twenty milliliters of the soil extract was dried under a stream of N_2, redissolved in 20 ml hexane, and loaded on the top of the column. Alkanes and biomarkers were eluted with 120 ml hexane; PAHs were eluted with 120 ml hexane and benzene (1:1). The eluates were dried under N_2, redissolved in 1 ml dichloromethane, and transferred to a vial for GC/mass spectrometry (MS) analysis. A mixture of internal standards was added (including acenaphthene-d12, chrysene-d12, naphthalene-d8, 1,4-dichlorobenzene-d4, perylene-d12, phenanthrene-d10) before analysis.

3.4.2.2 Gas Chromatography/Mass Spectrometry

The biomarkers and PAHs were analyzed by a Hewlett-Packard Model 6890 gas chromatograph equipped with a mass selective detector as described in Section 2.1. To achieve high sensitivity and low detection limits, specific ions were selected for each target analyte quantification by using selected ion monitoring (SIM). The quantitation ions are listed in Table 3.6. A recovery study was executed to ensure that the target compounds were quantitatively extracted from the soil, recovered from the chromatography column during separation, and properly analyzed by the GC/MS. Recoveries for typical compounds (phenanthrene, benzo[a]anthracene, pyrene, and benzo[a]pyrene) ranged from 97 to 103%.

3.4.3 Biomass Measurements

Aboveground biomass samples and root samples were taken during the project. Three of the six replicate plots were randomly selected for sampling. At each sampling time, three randomly selected points at least 0.7 m within the border of each plot were sampled. Biomass was clipped within a frame centered over the sample point. The data from the May 1996 and July 1996 samples were obtained using 0.09 m^2. For the September 1996 and October 1997 samples, 0.25 m^2 was used. These samples were dried for 48 hr at 60°C and weighed. Roots were sampled by extracting a 30-cm core using a soil sampler with an inside diameter of 5 cm. Soil cores were divided into three depths corresponding to 0 to 10, 10 to 20, and >20 cm

Table 3.6 Target Compounds and Biomarkers and
the Quantitation Ions Used in GC/MS SIM

Compound	Quantitation Ion
Biomarkers	
17a(H),21B(H)-30-Norhopane	191
17a(H),21B(H)-Hopane	191
Target compounds	
Phenanthrene	178
Pyrene	202
Chrysene	228
Benzo[a]anthracene	228
Benzo[a]pyrene	252
Benzo[e]pyrene	252
Internal standards	
Naphthalene-d8	136
Acenaphthene-d12	164
Chrysene-d12	240
Perylene-d12	264

belowground. There was variation in the size of the lowest core due to limited depth of the soil and penetration of the soil sampler. Roots were stored at 4°C until samples were processed.

Roots were extracted from the soil using repeated water rinses combined with agitation by hand and magnetic stirrer. Roots were recovered by passing the slurry through a mesh screen. After removing remaining debris from the cleaned roots, the roots were stained with methyl violet. Stained roots were prepared for scanning by spreading the wet roots on clear overhead transparencies and covering the samples with another transparency to form a slide. Up to three slides were prepared for samples with large root quantities. Roots were scanned using a water-shielded Hewlett-Packard 4C flat-bed scanner. Images were taken in black and white at a resolution of 300 dots per inch. Root images were analyzed for root length estimation, distribution of root diameter, root length density, and surface area density using Delta-T image-processing software. Scanned roots were dried and weighed to measure root mass.

3.4.4 Microbial Analyses

Field evaluations of bioremediation were monitored by microbial enumerations and microbial community assessment. Spread plate counts on tryptic soy agar (TSA) can be used to assess culturable colony-forming units (CFUs). Since not all petroleum degraders are culturable by known methods, an evaluation of microbial activity was needed. BIOLOG assay (BIOLOG, Inc., Hayward, CA) microtiter plates were used to assess the response of the microbial community to 95 different carbon substrates (Garland and Mills, 1991). A diverse community uses a greater number of substrates. Increased populations have greater intensities of substrate usage. The BIOLOG method was evaluated over time to determine the appropriate level of substrate usage to be recorded. The percent of most substrates had leveled off by the 72-hr reading without significant evaporation of the individual wells. Data were taken every 24 hr to compare to this trend, which was consistent. Total substrate use can be compared to the use of independent substrate groups. Most probable number (MPN) enrichment cultures were used to test for petroleum degraders. Although informative, the current method employed was quite laborious. By using the microbial evaluation methods to assess the community, further questions about the activity can be shown in the contaminant concentration, soil moisture data, and relative plant growth.

Soil was sampled to a depth of 2 ft (0.61 m) at the phytoremediation field site using a coring device. The core sample was removed and completely homogenized. One third of the sample was used for microbial characterization and one third was used for contaminant analysis. The remainder of the sample was archived in a designated freezer. Bacterial enumeration for these samples was conducted with a viable heterotrophic plate count by the spread plate method (Pepper et al., 1995). The technique is employed for the selection of fastidious microorganisms by using TSA (Difco). This medium was chosen to facilitate the growth of the largest range of microbes that are culturable. After assessing the soil moisture of the soil samples, approximately 10 g of wet soil was added to autoclaved 120-ml bottles. For the appropriate soil dispersal, 95 ml of 0.20% tetrasodium pyrophosphate was added to the soil, making a 10^{-1} dilution (Alef and Nannipieri, 1995). This dilution of soil was shaken on a rotary shaker at 150 rpm for 30 min. In a sterile hood, the soil solution was allowed to stand for 10 min for sediment settling. Serial dilutions were made using 1-ml aliquots that were transferred with a sterile 1-cc syringe to 9.5 ml of 0.85% sodium chloride. The suspensions were vortexed after each transfer to ensure proper mixing. The petri plates of TSA were inoculated with a subaliquot of 0.1 ml solution. The plates were surface spread with an alcohol flame-sterilized spreader (Wollum, 1982). Three replicates were performed per dilution per sample; the average value of the three replicates was used for the calculations. Once inoculated, the plates were inverted to prevent water from sitting on the plate surface and incubated at 25°C for 96 hr. The dilution counted had 30 to 300 CFUs per plate. The plates were counted at 24-hr intervals. As new colony growth declined and contamination was not present, the number of CFUs was recorded for group averages. Colony morphology was also observed at this time.

The BIOLOG (BIOLOG, Inc., Hayward, CA) assay for microbial community activity was modified from the intentional use for clinical identification of microorganisms. The Gram-negative BIOLOG plate was chosen for the type of organisms to be assessed by 95 different carbon substrates. The dried substrate, dye, and nutrients were present in each well prior to inoculation. The redox dye (tetrazolium) was present in each well to indicate the intensity of microbial respiration and the amount of substrate utilized.

The soil processing for the BIOLOG plates was performed in the same manner as for the plate counts. The dilution series ended at a 10^{-3} suspension to minimize the effects from soluble organic matter. Approximately 19 ml (two replicates at that dilution) was poured into a sterile petri dish. A repeating pipetter was used to transfer the 100-μl aliquots into the microtiter wells. The first well in the plate was a water blank for a control. The control was observed for changes that could be attributed to soluble organic material. The plates were incubated at 25°C for 96 hr and assessed every 24 hr. The absorbance at 590 nm was measured on a Titertik Multiskan Microplate Reader (Flow Laboratories, McLean, VA). Intensity readings that were greater than 40% of the control well reading were considered positive for substrate utilization. The plates were also visually inspected on a light table to determine consistent reading (binary data).

Petroleum degraders that are present in the indigenous population can be estimated with an enrichment culture. The MPN method is a simple dilution method of the soil in a Bushnell Haas broth (a mineral salts broth lacking a carbon source). The soil was diluted to 10^{-9} with five replicates of each dilution. A tetrazolium (redox) dye was added to each suspension and vortexed. The petroleum substrate was added to each suspension, and then each was capped with an aerated cap. Controls of the dye and diesel were included as precautions against contamination. Changes in the dye indicated microbial respiration and were recorded weekly until activity ceased. The number of positive responses at each dilution level was used for the final calculations. A table was used to determine the actual MNP of degraders (Alexander, 1982).

Table 3.7 Aboveground Biomass in May, July, and September 1996 and October 1997

Sample	g/m^2			
	May 1996	July 1996	September 1996	October 1997
Bermuda grass				
1d1	52.2	290.3	691.2	432
1d2	50.4	244.8	853.6	264
1d3	89.2	361.7	417.2	388
3d1	104.1	298.0	543.2	408
3d2	100.4	238.7	453.2	428
3d3	91.3	325.6	294.8	292
4d1	83.9	257.2	532.0	192
4d2	94.8	192.6	651.2	384
4d3	150.7	378.8	572.0	208
Mean ± SD	90.8 ± 28.0	287.5 ± 57.3	556.5 ± 154.5	332.9 ± 89.5
Tall fescue				
1c1	89.1	300.8	384.4	444
1c2	8.8	318.0	355.2	492
1c3	63.6	283.7	405.2	672
3c1	58.9	202.6	264.4	460
3c2	121.1	190.9		424
3c3	31.6	237.4	294.0	444
4c1	10.7	183.0	244.8	308
4c2	19.3	259.4	238.8	528
4c3	41.1	170.7	204.4	400
Mean ± SD	49.3 ± 35.7	238.5 ± 51.6	298.9 ± 69.3	463.6 ± 93.7
White clover				
1a1	5.8	140.8	229.2	0
1a2	4.4	114.8	246.4	0
1a3	0.0	193.1	183.6	0
3a1	1.8	124.0	224.4	0
3a2	24.3	139.3	182.0	0
3a3	63.7	128.3	156.0	0
4a1	0.0	66.8	170.4	0
4a2	43.2	44.1	188.8	0
4a3	4.4	136.3	173.2	0
Mean ± SD	16.4 ± 21.5	120.8 ± 41.0	194.9 ± 29.1	0

3.4.5 Waste Management

The soil samples shipped to KSU for analysis were stored in cold rooms and disposed of according to policies dictated by the KSU Department of Public Safety. All samples were logged in by KSU laboratory staff at the time of arrival and tracked until disposal. As shown in Table 3.7, the soil remaining in the study area had TPH levels below the state of Virginia's cleanup goal at the initiation of the field study (subsequently the goal was raised), as measured by U.S. Environmental Protection Agency– (EPA) certified laboratories.

3.5 RESULTS FROM FIELD STUDY

3.5.1 Plant Growth

All three vegetation treatments (Bermuda grass/rye grass, tall fescue, and white clover) established and grew well in the contaminated soil, but the pattern of growth varied for each treatment. Bermuda grass was established by sod and provided the most rapid cover and rooting for all species. Seeding perennial rye grass into the Bermuda grass in the fall helped provide winter growth and rooting while the Bermuda grass was dormant. By the 7-month vegetation sampling in May 1996, Bermuda grass/rye grass had the most root production of all treatments. Tall fescue also grew well throughout the study. Since it was established by seed, it took longer to establish a full root system. At the 12-month sampling in September 1996, tall fescue had the highest mass and density of roots for all treatments. Tall fescue also had a deeper rooting system than the other treatments. Both the tall fescue and Bermuda grass/rye grass persisted well through the relatively dry second growing season. Both treatments showed reduced growth in the second year, but maintained a complete canopy (Table 3.7).

White clover showed a different growth pattern than the grasses. Clover was slower to establish than the grasses, although it produced an excellent canopy by the end of 1 year of growth. Clover roots were observed for the presence of effective root nodules. Very few apparently effective nodules were found. High nitrogen fertilization rates could have suppressed nitrogen fixation. Although the clover was inoculated at planting, it is possible that a sufficient population of *Rhizobium* bacteria failed to establish in the soil. The white clover established and grew well through one growing season, but it did not thrive through the second growing season.

Under the conditions of this study, the biomass production, rooting, and vegetation persistence of the Bermuda grass/rye grass treatment and tall fescue were clearly superior to the white clover (Tables 3.8 to 3.10). In a long-term, low-maintenance situation, the grass treatments would be successful, while the clover treatment would be unsuccessful without additional management. It is likely that clover could have been maintained in the plot with supplemental irrigation under dry conditions. The clover was less competitive with weeds than were the grasses. The most effective way to maintain clover would probably be to combine it with a grass species.

3.5.2 Plant Growth and TPH Degradation

This study showed increased rates of TPH degradation in vegetated treatments compared to unvegetated control treatments (Section 3.5.6.1). At the final sampling time at 24 months, white clover showed the highest rate of TPH degradation followed by tall fescue and Bermuda grass/rye grass. Vegetated treatments had 9 to 19% more degradation than the unvegetated control. Since hydrocarbon degradation also took place in unvegetated treatments, it is this added effect of vegetation (perhaps called the phytoremediation effect) that must be examined to discriminate among the vegetation treatments to determine which plant

Table 3.8 Root Characteristics for Phytoremediation Study Area, May 1996

Vegetation	Depth (cm)	Root Length (mm)		Root Length Density (mm/ml)		Surface Length Density (mm²/ml)		Mean Root Diameter (mm)		Root Mass (g/m²)	
Bermuda	0–10	35,265[a]	(14,671)	160.9	(65.4)	174.2	(70.2)	0.35	(0.04)	110.1	(47.98)
	10–20	7,534	(5,491)	29.71	(21.8)	31.30	(23.5)	0.34	(0.05)	16.55	(9.91)
	>20	2,740	(2,598)	12.10	(12.0)	13.84	(13.6)	0.36	(0.06)	7.90	(5.86)
Fescue	0–10	15,514	(6,691)	80.64	(26.8)	102.7	(37.1)	0.38	(0.06)	53.32	(21.1)
	10–20	6,796	(2,558)	27.78	(7.87)	33.38	(9.72)	0.39	(0.03)	16.62	(6.79)
	>20	2,164	(1,010)	8.85	(2.96)	11.35	(4.11)	0.40	(0.03)	13.99	(17.4)
Clover	0–10	11,697	(374)	41.75	(9.40)	58.40	(11.7)	0.45	(0.01)	37.89	(2.57)
	10–20	1,931	(403)	8.24	(2.17)	9.38	(2.03)	0.37	(0.02)	2.27	(0.84)
	>20	673	(159)	2.41	(0.11)	2.48	(0.10)	0.33	(0.00)	1.58	(0.89)

[a] Mean (standard deviation).

Table 3.9   Root Characteristics for Phytoremediation Study Area, September 1996

Vegetation	Depth (cm)	Root Length (mm)		Root Length Density (mm/ml)		Surface Length Density (mm²/ml)		Mean Root Diameter (mm)	Root Mass (g/m²)	
Bermuda	0–10	36,985[a]	(16,393)	181.4	(94.7)	160.1	(87.8)	0.28 (0.03)	92.64	(56.6)
	10–20	9,859	(5,616)	37.1	(20.7)	31.8	(15.9)	0.28 (0.03)	28.09	(14.1)
	>20	2,497	(1,010)	14.0	(6.21)	13.47	(5.19)	0.30 (0.02)	5.56	(1.5)
Fescue	0–10	34,273	(10,269)	161.2	(76.1)	164.5	(73.2)	0.33 (0.02)	104.2	(56.0)
	10–20	19,588	(8,694)	75.8	(23.3)	73.3	(21.4)	0.31 (0.03)	49.62	(35.5)
	>20	9,087	(5,450)	32.9	(18.35)	34.1	(18.9)	0.33 (0.03)	21.0	(14.65)
Clover	0–10	20,323	(13,178)	59.9	(36.2)	72.6	(49.6)	0.36 (0.06)	111.6	(138.7)
	10–20	3,794	(1,577)	14.8	(5.57)	16.1	(6.90)	0.35 (0.05)	15.40	(7.33)
	>20	3,075	(5,185)	13.2	(13.7)	13.1	(14.0)	0.29 (0.05)	8.93	(16.1)

[a] Mean (standard deviation).

species treatments are superior. Among the vegetation treatments, white clover had the lowest root and aboveground biomass production, yet it had the highest rate of hydrocarbon degradation. Thus the relationship between plant growth and hydrocarbon degradation in this study did not follow the simple pattern that greater plant growth resulted in greater degradation rates. The white clover grew well in the first 12 months. Due primarily to dry conditions in the second year, some of the white clover died. Most of the roots had decomposed by the final sampling at 25 months. Although the tall fescue and Bermuda grass/rye grass continued to grow in the second year, these treatments also experienced significant root decomposition. In the first year of the study, it is possible that root growth and the increasing influence of plant roots on soil physical properties and microbial activity were important to hydrocarbon degradation, while in the second year, rates of root decomposition were more important.

While it is important to establish a healthy vegetation community in order to garner the beneficial effects of vegetation, other factors such as the rate of root turnover, root exudation patterns, and the influence of vegetation on soil physical properties appear to be just as important as the quantity of vegetation produced. The degradation of petroleum hydrocarbons results from the interaction of the contaminant, the soil physical properties, the microbial community, the plants, and the local climate conditions. While establishment and maintenance of vegetation are primarily a management tool, the plants likely have an indirect

Table 3.10   Root Characteristics for Phytoremediation Study Area, October 1997

Vegetation	Depth (cm)	Root Length (mm)		Root Length Density (mm/ml)		Surface Length Density (mm²/ml)		Mean Root Diameter (mm)		Root Mass (g/m²)	
Bermuda	0–10	18,929[a]	(5,564)	94.6	(27.8)	66.9	(19.7)	0.23	(0.02)	63.3	(20.8)
	10–20	3,580	(2,423)	17.9	(12.1)	16.0	(12.5)	0.27	(0.04)	9.5	(7.0)
	>20	1,860	(1,225)	9.3	(6.1)	9.7	(6.4)	0.33	(0.04)	6.6	(4.4)
Fescue	0–10	11,181	(4,362)	52.5	(21.8)	164.1	(17.7)	0.31	(0.09)	54.0	(26.8)
	10–20	3,904	(2,221)	19.5	(11.1)	18.9	(11.1)	0.31	(0.05)	13.1	(8.8)
	>20	4,763	(3,681)	23.8	(18.4)	25.9	(19.2)	0.36	(0.03)	19.9	(13.0)
Clover	0–10	2,043	(1,197)	10.2	(6.0)	9.3	(4.3)	0.31	(0.07)	8.2	(3.9)
	10–20	366	(214)	1.8	(1.1)	1.7	(0.9)	0.31	(0.02)	5.2	(7.5)
	>20	159	(50)	0.8	(0.2)	1.0	(0.5)	0.39	(0.08)	1.5	(0.6)

[a] Mean (standard deviation).

influence on contaminant degradation by creating an environment more favorable to dissipation of the contaminant.

3.5.3 Contaminant Uptake

The objective of this analysis was to determine the extent to which the PAHs accumulated in the roots and were translocated to the shoots. If the aboveground portions of the plants were accumulating these potential toxins at concentrations higher than background, then phytoremediation potentially could have a deleterious effect on the food chain. Two important challenges were faced in this analysis: (1) any soil residue from the plots that remained on the plant tissue would give a "false positive" for plant concentrations (this is a particular problem with roots that are difficult to clean completely) and (2) PAHs are ubiquitous in the atmosphere, and deposition on aboveground plant tissue has been observed even in remote, rural areas. Therefore, additional investigation was necessary to determine typical background concentrations and properly interpret the results.

In October 1997, representative root and shoot samples of fescue and Bermuda grass were removed from the plots, shipped to KSU under ice, extensively washed to remove residual soil (particularly from the roots), dried, and sent to A.D. Little (Cambridge, MA) for analysis of 41 PAHs (Figures 3.5 to 3.8). Due to the expense involved, only one sample of each tissue type was analyzed. Because nearly 100 g of dry biomass of each type was sent, the A.D. Little lab was able to attain detection limits below 10 µg/kg for all analytes.

The PAH profiles were similar for the shoot material of the two grasses. A significant number of PAHs (unsubstituted and those with side chains) were detectable. Substituted naphthalenes were present in the highest concentrations (<50 µg/kg except for C2-naphthalene with approximately 200 µg/kg). For the higher molecular weight compounds, perylene was the most abundant, with concentrations ranging from 50 to 100 µg/kg. The only difference in the profiles of the two grass species was that Bermuda grass had higher concentrations of all PAHs with molecular weights equal to or greater than fluoranthene. All compounds except perylene were 60 µg/kg.

The PAH profile for the roots of Bermuda grass was very similar to the profile for the Bermuda grass shoots. Phenanthrene concentration was higher in the roots and perylene was lower; otherwise, the roots and shoots were nearly identical. For fescue, the concentrations of all PAHs with more than two rings were much higher in the roots than the shoots. Nearly all exceeded 100 µg/kg and some exceeded 1000 µg/kg. These concentrations are typical for roots of plants growing in contaminated soils (Wild et al., 1992).

Uptake of PAHs has been studied for a wide variety of plant species. Concentrations of PAHs in crops were evaluated after the soil had been amended with sewage sludge (Wild et al., 1992). Adding sewage sludge for several years increased soil PAH content, and the PAHs persisted in the soils for many years (Wild and Jones, 1989). Carrots grown in soil amended with fresh sewage sludge accumulated PAHs in the outer layers of the roots (Wild and Jones, 1992), but only the low-molecular-weight compounds moved into the core, probably because of their moderate lipophilicity. However, PAH concentration in the carrot shoot was unaffected by the application of sludge with high PAH concentration.

Contamination of the aboveground portions of plants by PAHs normally is the result of atmospheric deposition (Wild et al., 1992), with the pollutants originating from automobile exhaust and industrial emissions (Kirchmann and Tengsved, 1991; Edwards, 1983, 1986; Jones, 1991). Benzo[a]pyrene concentrations in the air of urban areas of the U.S. range from 0.1 to 61.0 mg/m^3, but concentrations in nonurban areas range from only 0.01 to 1.9 mg/m^3 (Edwards, 1983). Airborne, low-molecular-weight PAHs are found predominantly in the vapor phase, whereas the multiringed compounds usually are associated with suspended

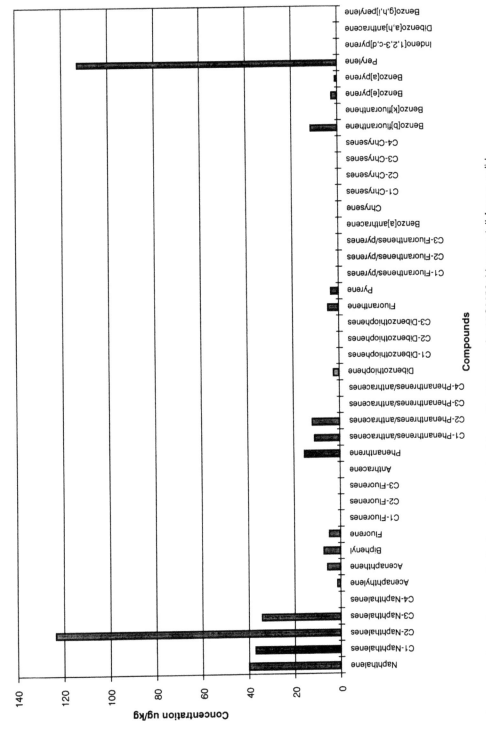

Figure 3.5 Contaminant uptake by fescue shoots. Analyte profile histogram for 97C3656: biomass tall fescue solid.

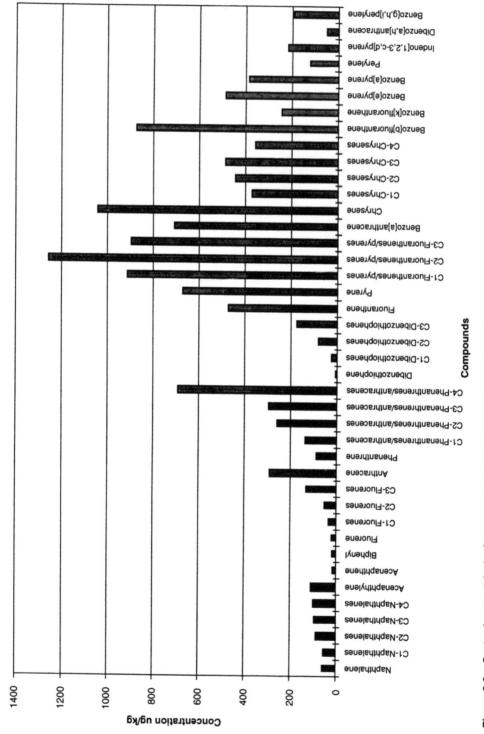

Figure 3.6 Contaminant uptake by fescue roots. Analyte profile histogram for 97C3657: root biomass tall fescue solid.

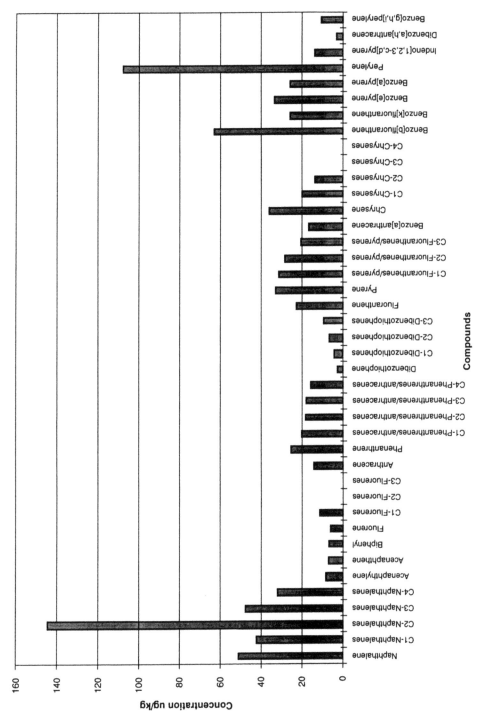

Figure 3.7 Contaminant uptake by Bermuda grass shoots. Analyte profile histogram for 97C3658: biomass Bermuda grass solid.

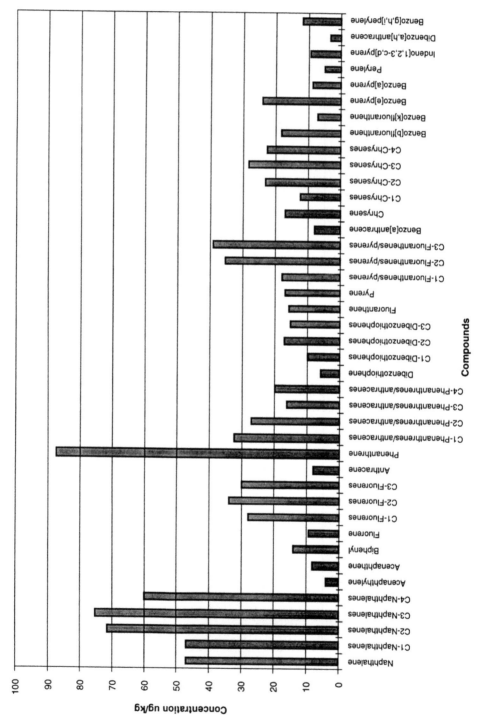

Figure 3.8 Contaminant uptake by Bermuda grass roots. Analyte profile histogram for 97C3659: root biomass Bermuda grass solid.

particulates. Those PAHs in the vapor phase can deposit on the leaf surface and become strongly adsorbed to the tissue. Sims and Overcash (1983) reported that plant shoots contained concentrations of individual PAHs from 22 to 88 μg/kg.

The vast majority of the mass of the PAHs is nearly always associated with the outer layer of cells of the roots, indicating strong adsorption to these cell walls, residual contamination of soil, or both. The very low concentrations of PAHs in the fescue shoots, despite the high concentrations in the roots, supported previous observations that only the low-molecular-weight compounds show any propensity for translocation from root to shoot. Therefore, phytoremediation of petroleum-contaminated soils should pose no threat to animals that graze on the plant shoots.

3.5.4 Leachate Analysis

Water is a critical component of the phytoremediation process, and irrigation is sometimes used to supplement natural rainfall during critical periods of the process. During this field demonstration, an aboveground, high-impact irrigation system was used to provide adequate water for seed germination and plant growth. Water was supplied to the system from a 6-in. water main at 60 psi and 200 gpm. A 2-in. PVC supply line carried water from the water main to the test plots. The site was divided into three zones to give complete overlap of coverage. Each zone had five Rainbird™ high-impact sprinkler heads attached to 3-in. risers. The site was irrigated several times daily for 5 min per zone during seed germination. After plant establishment, the site was irrigated as needed. Thirty minutes of operation resulted in the application of 1 in. of water over the entire study area. During the height of the summer, the plants required 2 to 3 in. of water per week for optimal growth.

Therefore, the possibility existed that some of the petroleum contaminants could be mobilized with the influent water and transported off-site. Leaching of petroleum products through the vadose zone is obviously undesirable, and one of the objectives was to determine the extent of leaching of contaminants, if any, at the demonstration site.

Vacuum pore-water samplers (lysimeters) were installed at several locations at the site. The lysimeters consisted of a porous ceramic cup sealed to the bottom end of a PVC tube. The top of the tube was sealed with a stopper equipped with an inlet and outlet port to allow the application of vacuum and withdrawal of samples. The lysimeters were buried in the soil (ceramic cup down) and covered with soil. Tubing ran from the stopper to the soil surface to allow sampling. Each vegetative treatment had at least two lysimeters, one at 30 cm depth and one at 50 cm. During each soil sampling event, the lysimeters were sampled by applying a vacuum for at least 2 hr, withdrawing the sample into a glass vial, shipping to KSU under ice, and storing for analysis. The lysimeters did not function if the soil was dry. Therefore, not every lysimeter yielded a sample during each sampling trip (Figure 3.9).

Lysimeter samples were analyzed using a modification of EPA Method 8015 in which C_{18} columns were used for sample cleanup and concentration prior to GC analysis. Ten-milliliter columns were packed with 2.5 g of C_{18}-coated silica and conditioned by eluting with 5 ml dichloromethane, 5 ml methanol, and 5 ml water. The lysimeter sample (up to 50 ml) was passed through the column, and all the hydrocarbons were retained on the column. The column was eluted with 2.5 ml of dichloromethane, water was removed with anhydrous Na_2SO_4, and the sample was sealed in a GC vial. The sample was analyzed for diesel-range TPH by GC/FID as described for soil samples. This procedure was extensively tested for recovery and reproducibility. It was found to be superior to liquid–liquid extraction and exceeded EPA quality assurance/quality control requirements.

All lysimeter samples were found to have no detectable TPH. For a 25-ml sample (the minimum volume), the detection limit was 1 mg TPH per liter. Undetectable TPH was

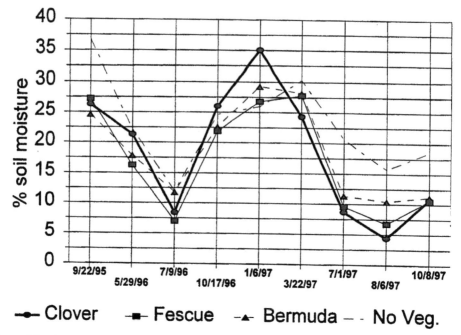

Figure 3.9   Average soil moisture (percent) in phytoremediation plots September 1995 to October 1997.

consistent with the visual observation of a lack of "sheen" on the surface of the samples often associated with petroleum contamination.

3.5.5 Microbial Characterization

The vegetated plots were planted in the late fall of 1995. By the spring of 1996, there were new roots developing for the first growing season. CFUs were evaluated for each soil treatment type at this time. In the fall of 1996, the overall numbers had fallen, except for the Bermuda treatment. By the summer of 1997, the CFUs for the unvegetated control and clover treatments had reached a plateau (Table 3.11). Microbial numbers in the Bermuda grass and fescue plots were similar to those found in the clover treatment until the fall of 1997. At that time, the Bermuda and fescue declined. Irrigation was absent for the summer of 1997, and soil moisture numbers declined (Figure 3.9). The biomass in the clover plots had declined due to water stress, and there was only a sparse invasion by other grass species. The established biomass in the fescue and Bermuda plots had extracted the soil moisture from the top 12 in., affecting microbial activity. This change in activity was further viewed in the BIOLOG assessment of the microbial community (Figure 3.10). Activity levels based on substrate use

Table 3.11   Soil Microbial Plate Counts in the Phytoremediation Study Area

	Log Microbial Numbers (CFUs/g Dry Soil)						
Treatment	March 1996	October 1996	January 1997	March 1997	May 1997	July 1997	October 1997
Clover	8.81	7.99	6.21	7.11	6.56	6.92	6.73
Fescue	8.60	7.62	6.54	6.97	6.42	6.79	5.77
Bermuda	7.36	7.99	6.05	7.26	6.58	6.76	5.91
Unvegetated control	8.24	7.17	5.63	6.79	6.89	6.91	6.83

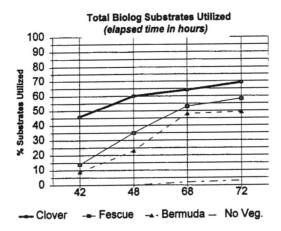

Figure 3.10   BIOLOG substrate utilized over time.

were consistently higher for vegetated treatments in the 1996 growing season. Without the continued irrigation in the summer of 1997, microbial activity decreased gradually over the growing season.

Petroleum degraders were enumerated in the spring and fall of 1997, the second growing season. The MPN technique estimated that the highest amount of degraders was present in the clover and fescue plots in the spring (Table 3.12). The Bermuda grass is a warm-season plant that does not have high activity in the spring. A late-season sampling in the fall of 1997 indicated that the number of petroleum degraders was higher in the vegetated plots than the unvegetated control, with clover having the highest number. The clover plots also had the lowest TPH concentration of all the plots at this sampling time.

Data showing utilization of carbon sources grouped by chemical structure indicated that there were also differences between type of substrate utilized by microbes from different planted plots (Figures 3.11 and 3.12). Microbial communities found in the clover plots generally utilized the highest percentage of aromatics, carboxylic and amino acids. Results from the Bermuda and fescue plots were similar, with relatively high substrate utilization after 68 hr of incubation. Microbial communities from the unvegetated plots utilized very few of these compounds efficiently. In year 2, the BIOLOG differences between plots were not significant.

3.5.6 Contaminant Dissipation

3.5.6.1 TPH Data

The soil samples were analyzed after drying and grinding, as discussed in Chapter 2. For quality assurance/quality control, all EPA protocols were followed in terms of instrument

Table 3.12   MPN of Petroleum Hydrocarbon Degraders in Soil in the Phytoremediation Study Area

Treatment	March 1997	October 1997
Clover	5.4×10^7	3.9×10^7
Fescue	3.4×10^7	1.2×10^6
Bermuda	4.3×10^6	1.3×10^6
Unvegetated control	6.5×10^6	3.0×10^5

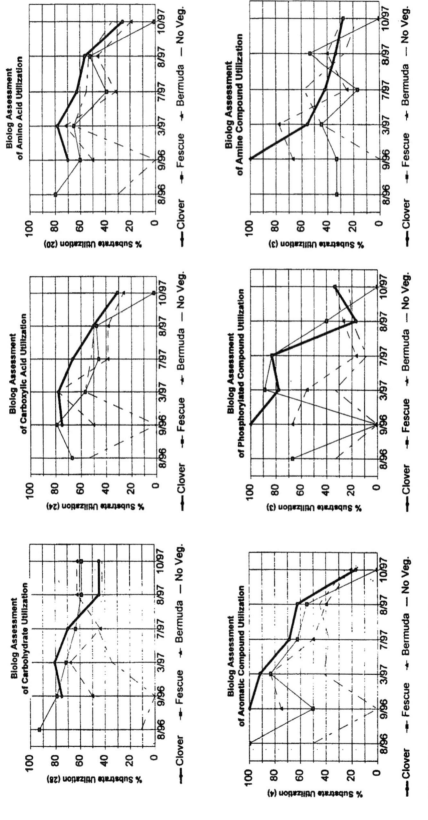

Figure 3.11   BIOLOG assessment using suites of compounds I.

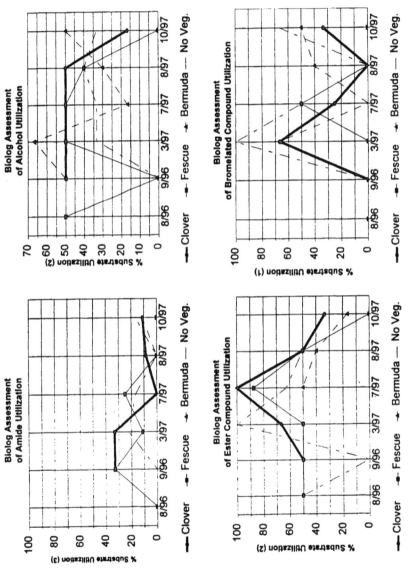

Figure 3.12   BIOLOG assessment using suites of compounds II.

Table 3.13   TPH Concentration in Soil During the Phytoremediation Field Demonstration

	mg/kg									
Treatment	November 1995	March 1996	May 1996	July 1996	September 1996	November 1996	March 1997	May 1997	July 1997	October 1997
Clover										
Replicate 1	2573	2435	2387	2057	1856	1839	1775	1523	2002	1289
2	3268	3060	3087	2438	2315	1859	1869	1943	1781	1610
3	2260	1864	1980	1869	1729	1598	1585	1546	1341	1176
4	2293	2294	2358	1924	1967	1743	1667	1418	1576	1099
5	1925	1821	1558	1708	1175	1323	1420	1488	1329	1023
6	1677	1430	1261	1640	1492	1348	1319	1463	1083	829
Unvegetated										
Replicate 1	2172	2059	1441	1811	1526	1907	1631	1649	1685	1372
2	2443	2146	1923	2042	2145	1915	1791	1713	1833	1819
3	2244	2063	1769	1950	1993	1680	1811	2014	1729	1564
4	1786	1984	1880	1834	1662	1452	1552	1689	1190	1274
5	1646	1781	1348	1711	1464	1289	1332	1466	1182	1145
6	2020	1883	1448	1729	1574	1424	1518	1625	1149	1312
Fescue										
Replicate 1	2024	1851	1797	1803	1901	1520	1597	1496	1729	1204
2	3015	2964	2914	3040	2677	2050	2233	1717	1252	1537
3	1813	1973	2174	2067	1705	1490	1591	1599	1332	1063
4	2270	1846	1906	1524	1534	1320	1174	1260	940	1147
5	2290	2064	1526	1528	1479	1213	1213	1238	1212	1179
6	2017	2039	1775	1821	1521	1353	1375	1375	1090	1203
Bermuda										
Replicate 1	2590	2868	2252	2345	2280	1934	1974	2028	1323	1720
2	2357	2115	2129	2027	1757	1475	1557	1540	1885	1559
3	1958	2053	1781	1877	1827	1390	1537	1276	1307	1197
4	2089	1858	1778	1844	1580	1597	1262	1214	1234	1243
5	2153	1821	1372	2059	1615	1543	1489	1408	1288	1086
6	1971	1884	2099	2134	1760	1583	1720	1690	1271	1169

performance, surrogate spike, internal standards, and replication. The instrument performance standard was run before each set of analyses, and the results fell within suggested criteria each time. Recoveries of surrogate spike averaged 93%. If recoveries or reproducibility of replicate analyses fell below EPA limits, the samples were run again. This criterion was upheld rigidly. Individual samples were frequently rerun, and an entire set of samples had to be reanalyzed near the end of the experiment. Reanalyzed samples were strictly held to the quality assurance/quality control criteria.

Soil contaminant concentrations are shown in Tables 3.13 to 3.15 and Figure 3.13 for most sampling times. Statistical analysis of these data is described in Section 3.5.6.3. The vegetated plots had statistically significant higher percentages of TPH degradation than the unvegetated control plots. All percentages of degradation increased significantly throughout the project, including the unvegetated plots. The Bermuda grass plots were significantly lower than the fescue plots. After 24 months of plant growth, the plots containing plants had significantly lower soil TPH concentrations than the unvegetated control plots.

When the TPH degradation data were considered, the white clover treatment performed better than the grasses. An important conclusion of this study was that plant species that provide the highest rate of petroleum hydrocarbon degradation may not be the species with the highest growth rates or the species that perform the best under low-maintenance conditions (see Section 3.5.1). The relationship between plant growth and petroleum hydrocarbon degradation should continue to be monitored in future studies.

Table 3.14   Percent Reduction in Soil TPH Concentration in the Phytoremediation Field
Demonstration

					%				
Treatment	March 1996	May 1996	July 1996	September 1996	November 1996	March 1997	May 1997	July 1997	October 1997
Clover									
Replicate 1	5	7	20	28	29	31	41	22	50
2	6	6	25	29	43	43	41	45	51
3	18	12	17	23	29	30	32	41	48
4	0	−3	16	14	24	27	38	31	52
5	5	19	11	39	31	26	23	31	47
6	15	25	2	11	20	21	13	35	51
Unvegetated									
Replicate 1	5	34	17	30	12	25	24	22	37
2	12	21	16	12	22	27	30	25	26
3	8	21	13	11	25	19	10	23	30
4	−11	−5	−3	7	19	13	5	33	29
5	−8	18	−4	11	22	19	11	28	30
6	7	28	14	22	30	25	20	43	35
Fescue									
Replicate 1	9	11	11	6	25	21	26	15	41
2	2	3	−1	11	32	26	43	58	49
3	−9	−20	−14	6	18	12	12	27	41
4	19	16	33	32	42	48	44	59	49
5	10	33	33	35	47	47	46	47	49
6	−1	12	10	25	33	32	32	46	40
Bermuda									
Replicate 1	−11	13	9	12	25	24	22	49	34
2	10	10	14	25	37	34	35	20	34
3	−5	9	4	7	29	22	35	33	39
4	11	15	12	24	24	40	42	41	41
5	15	36	4	25	28	31	35	40	50
6	4	−7	−8	11	20	13	14	35	41

3.5.6.2 Target Compounds

When analyzing the suite of compounds found in petroleum and related products, one can identify and quantify specific saturated hydrocarbons and PAHs for the study of the weathering and fate of petroleum in the environment. The distribution of PAHs and biomarkers can be used for contaminant source identification or as a measure of the extent of degradation (Douglas et al., 1996).

Table 3.15   Statistical Analysis of Average Percent Reduction in TPH in the Phytoremediation
Field Demonstration

Treatment	March 1996	May 1996	July 1996	September 1996	November 1996	March 1997	May 1997	July 1997	October 1997
Clover	8	11	15	24	29	30	31	34	50
Unvegetated	2	20	9	16	21	21	17	29	31
Fescue	5	9	12	19	33	31	34	42	45
Bermuda	4	13	6	17	27	27	30	36	40
LSD (0.05)	ns	ns	ns	ns	9	12	13	15	6
LSD (0.1)	ns	ns	ns	ns	8	11	11	12	5

Note: LSD = least significant difference at a probability of significance of either 0.05 or 0.1. ns = no significant difference.

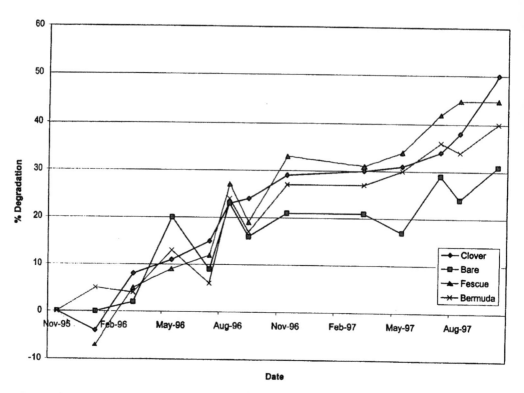

Figure 3.13   TPH degradation over time in phytoremediation plots.

The selected ion-monitoring mode used in this analysis has excellent sensitivity because only a few of the possible ions are measured, allowing acquisition of more of the ions during the run. Certain molecular ions are common to many analytes (e.g., 191 m/z for biomarkers) with the remaining fragments depending upon the initial molecular mass.

Two hopanes were chosen as biomarkers for this study because they are relatively recalcitrant and easily quantified in these samples. The two compounds were readily isolated from each other and from other compounds by GC. The compounds were successfully identified through MS by comparing appropriate ion ratios to library spectra. During the period from October 1995 to November 1996, the changes in the concentrations of the biomarkers were less than 2%, as compared to more than 30% for the changes in TPH during the same period (Table 3.16). Thus, these hopanes appear to be satisfactory "biomarkers."

Changes in the PAHs over the 13-month experimental period are illustrated in Table 3.17. For all compounds except chrysene and benzo[e]pyrene, degradation was greater in the presence of white clover than in the absence of plants. A similar comparison could be made for tall fescue vs. the unvegetated treatments. Although the mean values for degradation in the presence of fescue were always greater than those for the unvegetated plots, the difference was not always statistically significant. Nevertheless, there was a clear trend that, as with TPH changes, the presence of clover and fescue enhanced the dissipation of PAHs relative to bioremediation in the absence of plants. The behavior of chrysene over this time period was different from the other PAHs not only because it showed no differences due to treatment but also because rates of degradation were small (6 to 11%). This result is consistent with the observations of Douglas et al. (1994), who reported less than 20% degradation of chrysene after 16 months compared to 36% for pyrene.

Table 3.16   TPH and Biomarker Concentrations in Soil After 1 Year of Phytoremediation

	October 1995 Mean (SD), $n = 6$ (mg/kg)		November 1996 Mean (SD), $n = 6$ (mg/kg)		Ratio[a] (%)
TPH					
White clover	2988	(901)ab[b]	1929	(346)b	34.6
Tall fescue	2877	(353)a	1913	(438)b	32.9
Unvegetated	2799	(501)a	1987	(430)b	27.5
17a(H),21b(H)-Norhopane					
White clover	1.402	(0.499)a	1.400	(0.292)a	0.14
Tall fescue	1.355	(0.222)a	1.334	(0.154)a	1.55
Unvegetated	1.291	(0.200)a	1.262	(0.222)a	2.25
17a(H),21b(H)-Hopane					
White clover	1.788	(0.621)a	1.765	(0.190)a	1.29
Tall fescue	1.711	(0.263)a	1.696	(0.272)a	0.88
Unvegetated	1.640	(0.225)a	1.608	(0.290)a	1.95

[a] Ratio for TPH = $\{[(C_0/H_0) - (C_1/H_1)]/(C_0/H_1)\} \times 100$, where H_0, C_0 = October 1995 mean and H_1, C_1 = November 1996 mean. Ratio for biomarkers = $[(H_0 - H_1)/H_0] \times 100$; H_0 = October 1995 concentration and H_1 = November 1996 concentration.

[b] Means with the same letter are not significantly different ($p < 0.05$).

Table 3.17   PAH Concentration in Soil After the First Year of the Phytoremediation Field Demonstration

	October 1995 Mean (SD), $n = 5$ (mg/kg)		November 1996 Mean (SD), $n = 5$ (mg/kg)		Ratio[a] (%)
Phenanthrene					
White clover	0.844	(0.139)ab[b]	0.695	(0.253)ab	17.6
Tall fescue	1.182	(0.787)a	1.04	(0.347)ab	12.0
Unvegetated	0.581	(0.492)b	0.566	(0.213)b	2.58
Pyrene					
White clover	1.099	(0.214)b	0.579	(0.375)c	47.3
Tall fescue	1.692	(0.471)a	0.821	(0.460)b	51.5
Unvegetated	1.050	(0.275)bc	0.873	(0.457)bc	26.2
Chrysene					
White clover	0.417	(0.031)ab	0.393	(0.052)b	5.76
Tall fescue	0.492	(0.094)a	0.438	(0.099)ab	10.9
Unvegetated	0.450	(0.063)ab	0.419	(0.100)ab	6.89
Benzo[a]anthracene					
White clover	0.624	(0.039)b	0.417	(0.108)c	33.2
Tall fescue	0.846	(0.219)a	0.556	(0.191)bc	34.3
Unvegetated	0.657	(0.083)b	0.543	(0.155)bc	17.4
Benzo[a]pyrene					
White clover	0.777	(0.145)a	0.661	(0.023)a	14.9
Tall fescue	0.759	(0.082)a	0.711	(0.160)a	6.30
Unvegetated	0.786	(0.087)a	0.780	(0.124)a	0.76
Benzo[e]pyrene					
White clover	0.653	(0.095)ab	0.522	(0.050)b	20.6
Tall fescue	0.746	(0.192)a	0.610	(0.194)ab	18.2
Unvegetated	0.661	(0.096)ab	0.567	(0.129)b	14.2

[a] Ratio of PAH = $[(C_0 - C_1)/C_0] \times 100$, where C_0 = October 1995 mean and C_1 = November 1996 mean.

[b] Means with the same letter are not significantly different ($p < 0.05$).

Table 3.18   Degradation of Selected PAHs, as Determined by A.D. Little

	% Degradation				
Treatment	C2- Phenanthrene	C3- Phenanthrene	C4- Phenanthrene	C1- Dibenzothiophene	C2- Dibenzothiophene
Clover	59a	68a	69a	61a	76a
Unvegetated	26b	31b	35ab	19b	46b
Fescue	62a	61a	40ab	64a	70a
Bermuda	56a	52ab	13b	48ab	49ab

	% Degradation			
Treatment	C3- Dibenzothiophene	Fluoranthene	Pyrene	C1- Fluoranthene
Clover	77a	38b	54ab	36ab
Unvegetated	42b	51b	47b	27b
Fescue	68ab	86a	71a	53a
Bermuda	59ab	42b	44b	37ab

Note: Values with the same letters are not significantly different.

Soil samples from November 1995 and October 1997 were analyzed by A.D. Little for PAHs. Analytical difficulties were encountered with a few of the samples and required that one replicate be dropped from the statistical analysis. For the lower and higher molecular weight compounds, a problem of unknown origin resulted in large increases in concentrations from the initial to the final sampling period. Because increases in PAH concentrations were not likely and increases of 1500% could not be rationalized, the results for only 10 of the compounds are reported in Table 3.18.

Degradation in the unvegetated treatment was significantly less than that in at least one vegetated treatment except for C4-phenanthrene. For all compounds except fluoranthene, the degradation in the clover treatment was the greatest or statistically equivalent to the greatest. The Bermuda grass treatment frequently had the least degradation among the vegetated plots. The results in Table 3.18 were consistent with the laboratory TPH analyses and PAH concentrations determined for the 13-month sampling.

In an attempt to remove heterogeneity from the PAH concentration data, the concentrations were normalized to an oil-weight basis by A.D. Little. To accomplish this, the gravimetrically determined extract weight (post alumina cleanup) was substituted for the soil weight in the concentration calculation. The concentration was thus the concentration of the particular PAH in the extracted oil, not that of the soil. Not only did this reduce the effects of heterogeneous TPH concentrations, but it also allowed evaluation of the degradation of the actual oil components.

The trends in the PAH data were very consistent (Table 3.19). For nearly every compound, the percent degradation was the greatest, or statistically equivalent to the greatest, in the fescue plots. The vegetated treatments significantly outperformed the unvegetated control with the exception of the C2-, C3-, and C4-naphthalenes; biphenyl; C4-phenanthrene/anthracene; benzo[b]- and -[k]-fluoranthene; and perylene. For some compounds, such as the chrysenes, the percent degradation in the clover plots was statistically less than in the fescue and equivalent to that in the unvegetated plots.

Apparently, the trends observed in the TPH concentrations could not be directly transferred to PAH concentrations. The clover plots had significantly greater TPH dissipation than all other plots, yet had one of the poorest performances for PAH dissipation. Degradation of both TPH and PAH in the fescue plots was among the best. In the clover treatments, degradation of TPH increased sharply during the period when the clover was dying and the

Table 3.19 PAH Degradation Measured Using Oil-Based Weight Method by A.D. Little

Compound	White Clover	Unvegetated Control	Tall Fescue	Bermuda Grass	LSD 0.05 (0.1)
C1-Naphthalenes	1.4	30	29	−15	ns
C2-Naphthalenes	53ab	58ab	49ab	8b	48
C3-Naphthalenes	80a	73ab	71ab	49b	27
C4-Naphthalenes	87a	86a	71ab	61b	19
Acenaphthylene	−60	−22	36	7	ns
Acenaphthene	66	83	80	68	ns
Biphenyl	−24b	15ab	22ab	37a	57
Fluorene	62	75	75	62	ns
C1-Fluorenes	58b	65b	100a	84ab	30
C2-Fluorenes	84ab	68b	100a	87ab	32
C3-Fluorenes	91ab	71ab	100a	85ab	28
Anthracene	32c	42bc	65a	53ab	17
Phenanthrene	58	63	75	54	ns
C1-Phenanthrenes/anthracenes	50(ab)	41(b)	80(a)	63(ab)	ns (33)
C2-Phenanthrenes/anthracenes	74	62	86	70	ns
C3-Phenanthrenes/anthracenes	82(ab)	68(b)	86(a)	76(ab)	ns (16)
C4-Phenanthrenes/anthracenes	85a	74ab	81a	63b	14
Dibenzothiophene	46	59	72	53	ns
C1-Dibenzothiophenes	77	60	80	57	ns
C2-Dibenzothiophenes	85(ab)	65(b)	88(a)	73ab	ns (20)
C3-Dibenzothiophenes	88a	66b	89a	79ab	18
Fluoranthene	79b	80b	93a	80b	6
Pyrene	82ab	72b	93a	75ab	14
C1-Fluoranthenes/pyrenes	73ab	62b	84a	70ab	16
C2-Fluoranthenes/pyrenes	58ab	52b	69a	56ab	15
C3-Fluoranthenes/pyrenes	50ab	36c	59a	45bc	13
Benzo[a]anthracene	71b	61b	86a	64b	13
Chrysene	60b	55b	78a	54b	13
C1-Chrysenes	59bc	53c	74a	68ab	15
C2-Chrysenes	54ab	44b	65a	60a	14
C3-Chrysenes	23b	28b	50a	49a	20
C4-Chrysenes	25	26	37	27	ns
Benzo[b]fluoranthene	43b	59ab	67a	49b	18
Benzo[k]fluoranthene	30(b)	48(ab)	53(a)	45(ab)	ns (20)
Benzo[e]pyrene	32	30	37	27	ns
Benzo[a]pyrene	49b	51b	69a	53b	15
Perylene	−8b	26a	25a	−3b	19
Indeno[1,2,3,-c,d]pyrene	−42	14	30	22	ns
Dibenzo[a,h]anthracene	37	35	43	45	ns
Benzo[g,h,i]perylene	−35	19	35	22	ns

Note: Means with the same letter are not significantly different at $p = 0.05$ (0.1). LSD = least significant difference.

roots were degrading. The hydrocarbons possibly were degraded cometabolically, which impacted the more labile compounds first and the recalcitrant PAHs last. The coarse root structure of the clover had limited capability to explore the microsites in the soil where PAHs strongly adsorb and avoid degradation. Fescue, on the other hand, had a much finer root structure that penetrated a greater volume and smaller sites in the soil than clover and, therefore, promoted the degradation of a greater percentage of the PAHs.

3.5.6.3 Statistical Analyses

The field design was a randomized complete block with four treatments and six blocks (replicates). The response measured was TPH concentration. Analysis of variance (ANOVA) was run on the initial concentrations as a simple test of homogeneity (COSTAT, CoHort

Software, Berkeley, CA). The concentrations as a function of block were found to be significant, but there were no differences among the means within the vegetation treatments. This suggested that (1) the original design of complete blocks was appropriate and (2) covariate analysis of the final data may be necessary to determine if differences in initial concentrations could have impacted the rate of degradation. Analysis of covariance was used in addition to standard ANOVA to add to the power of the statistical test (SAS, SAS Institute, Cary, NC).

Analysis of Variance — The ANOVA yielded the mean separation test (Duncan's multiple range, $p < 0.05$) used to determine the statistical differences among the means at each sampling time. Because the change in concentration was of interest (and because initial starting concentrations varied significantly across blocks), percent change was used as the variable.
   Percent change is defined as:

$$\% \text{ change} = 100 \times \frac{(\text{initial concentration} - \text{concentration at time } t)}{\text{initial concentration}}$$

This transformation normalized the concentrations and removed the effects of the initial concentrations. (This simple approach accounted for the same effects as covariate analysis but was not as robust.) Values of "mean % change" at a given time followed by the same letter are not statistically different.

Analysis of Covariance — A simple explanation of covariate analysis is that this statistical approach compensates for differences in starting values by assuming that they can be adjusted to a central value using linear regression. Standard ANOVA of a randomized complete block experiment assumes that the value of a given observation, Y_{ij}, may be expressed as

$$Y_{ij} = \mu + R_i + C_j + \varepsilon_{ij}$$

where μ is the true mean, R_i is the block effect, C_j is treatment effect, and ε_{ij} is the error.
   In covariate analysis, the model is somewhat different:

$$Y_{ij} = \mu + R_i + C_i + \beta(x_{ij} - x[\text{mean}]) + \varepsilon_{ij}$$

in which $\beta[x_{ij} - x(\text{mean})]$ is the covariate correction. The values of R_i and C_j are not the same as in ANOVA.
   In the field experiment at Craney Island, the simplicity of the experimental design made the analysis of covariance straightforward. However, there were several possible responses to analyze, such as TPH, change in TPH, or relative change in TPH as percent reduction. Several options were pursued, and each is discussed below. However, due to the complexity of the statistical analysis, only the final sampling data were analyzed in this manner (Table 3.20).

Residual TPH — The remaining TPH in the soil was analyzed using analysis of covariance to determine if initial concentration had an impact on rate of degradation and to assist in determining the level of rigor to be used in discussing the final concentrations as affected by vegetation treatment.
   The SAS output indicated that the impact of cover was highly significant ($p < 0.0005$) and final values were impacted significantly by the initial values. After correction for covariance,

Table 3.20 Mean Separation Resulting from the Analysis of Covariance Using the Initial and Final TPH Concentrations

Treatment	Residual TPH	Change in TPH	% Change
White clover	1124a[a]	1077a	49a
Tall fescue	1209a	992a	45ab
Bermuda grass	1334b	867b	40b
Unvegetated	1468c	734c	32c

[a] Note that all values have been "corrected" through covariate analysis. Values with the same letter are not significantly different.

the effect of blocks was not significant. "Corrected" treatment TPH means followed the same general trends observed for ANOVA. Mean TPH concentration in the plots with clover and fescue was significantly less than in the plots with Bermuda, which was less than unvegetated. The TPH in clover and fescue was approximately 250 to 300 mg/kg less than in the unvegetated treatment; thus, the differences were not only statistically different but pragmatically important as well.

Change and Percent Change in TPH — The change in TPH concentrations followed the same trends as the residual TPH values. Although this was expected intuitively, the impact of the correction for covariance is sometimes difficult to predict. The relative change (percent change) in TPH followed the same general trend. All vegetative treatments were greater than unvegetated, but changes in the fescue and Bermuda treatments were not significantly different.

The analysis of covariance produced essentially identical results as ANOVA, which illustrates how initial concentrations may have had an impact on the degradation of TPH but did not affect the final results. Thus, one still can state with confidence that the presence of all plant species used in this study significantly increased degradation of TPH compared to plots with no species but otherwise treated identically.

Percent Changes in PAH Concentrations — Due to the extreme cost in materials and labor, soil concentrations of selected PAHs were determined on only five of the six blocks, three of the four vegetated treatments (Bermuda was excluded), two sampling times, and only one analysis per block. As with TPH, ANOVA was performed on the percent change in concentration. However, due to the greater variability in the data (because of fewer samples and more steps in the chemical analysis), the data were transformed prior to running the ANOVA. Mean separation was determined for $p < 0.05$.

3.5.6.4 Independent Laboratory Results

One soil sample from each treatment in each block (24 total samples) was sent to an outside analytical laboratory for TPH analysis in October 1995, September 1996, and September 1997. The laboratories used soxhlet extraction of field-moist samples followed by GC/FID analysis. The 1995 and 1996 samples were handled by the same laboratory as used by the Navy for analysis of the biotreatment facility samples, but it was necessary to find another analytical lab when the first went out of business in 1996.

The outside laboratory's analysis of the 1995 samples (Table 3.21) found considerably less TPH than determined by any of the KSU procedures. The difference between the results cannot be definitely explained without rigorous interlaboratory comparisons. However, the drying and grinding procedures used in the KSU laboratory may have played an important

PHYTOREMEDIATION OF HYDROCARBON-CONTAMINATED SOIL

Table 3.21 Independent Laboratory Results for Soil TPH Concentration in
            Phytoremediation Study Area

Treatment	October 1995[a] (mg/kg)	September 1996[a] (mg/kg)	September 1997[b] (mg/kg)
White clover			
Replicate 1	723	230	140
2	1030	176	530
3	1290	114	140
4	509	59	250
5	1160	314	130
6	484	565	200
Unvegetated			
Replicate 1	1140	230	50
2	6620	<20	610
3	1230	53	350
4	1100	<20	250
5	697	98	260
6	575	<10	210
Tall fescue			
Replicate 1	885	575	100
2	992	177	490
3	1270	94	300
4	463	89	320
5	556	21	210
6	1320	33	200
Bermuda grass			
Replicate 1	1380	77	490
2	882	44	660
3	1270	38	370
4	1070	209	210
5	534	90	340
6	860	1070	200

[a] Environmental Testing Services, Norfolk, VA.
[b] Continental Testing Services, Salina, KS.

role in the differences. The KSU laboratory studies found that a wet sample was not extracted efficiently by soxhlet. These suspicions were supported by the external lab's results in 1996. Its reported TPH concentrations generally declined from 1995 to 1996, but the biggest declines were observed in the unvegetated plots (contrary to the KSU determinations in which the least decline was observed in the unvegetated plots). When KSU collected samples in September 1996, the field was very moist, particularly in the unvegetated plots in which there were no active plants to remove the excess water. Therefore, when the laboratory extracted the samples without prior drying, the negative impact of residual moisture would have been greatest in the samples from the unvegetated plots.

The analyses from September 1997 were performed by the second outside laboratory. The samples were not as moist as in 1996, which probably contributed to the increase in reported concentrations from 1996 to 1997. Also, direct comparisons between laboratories are not always possible.

Along with the shaking extraction of dried, ground samples, KSU extracted field-moist samples from three dates by soxhlet (Table 3.22). At all three sampling times, the concentrations from the soxhlet extracts were lower than from the shaking; however, the concentrations did not decrease as did the concentrations from shaking extracts. The differences in concentrations can be attributed to the effect of moisture and contaminant aging on soxhlet extraction and subsampling variability due to the moist soils. Extraction of moist soils was

Table 3.22 TPH Concentration as Measured by Wet Soxhlet Extraction with GC/FID at KSU

Treatment	November 1995 (mg/kg)	August 1996 (mg/kg)	October 1997 (mg/kg)
White clover			
Replicate 1	999	978	1111
2	1237	1463	973
3	717	1201	910
4	1054	1235	860
5	682	926	835
6	755	799	699
Unvegetated			
Replicate 1	487	1103	757
2	690	1155	674
3	474	904	967
4	1310	1045	845
5	793	934	694
6	856	950	653
Tall fescue			
Replicate 1	1080	1460	1094
2	761	773	1467
3	444	1168	1224
4	638	1086	927
5	578	455	961
6	674	773	805
Bermuda grass			
Replicate 1	676	743	1171
2	892	1001	1039
3	424	360	1052
4	594	884	897
5	394	282	854
6	572	298	1107

found earlier to have greater variability than dry soil (Table 2.2), and aged contaminants were not extracted efficiently by soxhlet (refer to Tables 2.5 and 2.6).

The soxhlet procedure used by the outside laboratories and KSU yielded lower TPH concentrations than the shaking extraction method developed for this project, and this was no surprise. It was suspected from the beginning that a passive extraction would have less opportunity to remove entrapped contaminants, and the data support this hypothesis. The Craney Island samples extracted and discussed in Section 2.1 were given the best opportunity for higher values by soxhlet because they were extensively homogenized (sieving, mixing) and were drier than any of the subsequent samples (the samples described in Section 2.1 were obtained prior to plot establishment). Despite the fact that soxhlet is the regulatory standard, the data in this study show it to be inadequate unless the sample is nearly dry (not simply mixed with anhydrous sodium sulfate) and sieved to physically destroy soil aggregates.

ANOVA of the changes and percent change in TPH revealed no significant differences due to treatment. The mean change ranged from 634 to 605 mg/kg, with unvegetated having the smallest change. However, the least significant difference (LSD) was 405 mg/kg. Transforming the data to percent change did not improve the analysis; the range was from 69 to 59% with an LSD of 23%. The most important aspects of these data are that the TPH concentrations declined from 1995 to 1997 and the 1997 values were below the 700-mg/kg regulatory limit established by the state of Virginia for the Craney Island bioremediation cell.

3.6 SITE MANAGEMENT ISSUES

During the 1996 growing season, water was applied to the site via irrigation. This resulted in healthy microbial communities and vegetation. Both microbial numbers and plant biomass decreased during the 1997 growing season when irrigation was not possible because of a lack of on-site supervision and theft of irrigation parts. The lack of irrigation was compounded by a rather severe drought. It is probable that degradation rates were driven by water availability during the summer months of 1997, with an increase in reduction in the early fall due to increased precipitation. The importance of water content was confirmed in concurrent green-house studies conducted at KSU (Section 2.3). Phytoremediation of PAHs and diesel fuel over depth displayed a strong tie to available water.

The only amendments added to the site during the phytoremediation study were nitrogen and phosphorus. Both plants and microorganisms require adequate nutrients for healthy growth. The rate of fertilization was based on a 100:20:10 ratio of C:N:P, which is commonly used by environmental engineers to assess required nutrients. Periodic soil testing showed that available N and P increased in the unvegetated plots, causing a slight drop in pH (Tables 3.1 to 3.3). The pH remained constant in the vegetated plots, as did the levels of available N and P, showing that the vegetation was capable of using all of the applied fertilizer.

Based on the reduced degradation rates experienced during the 1997 growing season, active site management is a necessary component for the success of phytoremediation. Many environmental factors, such as temperature, wind, and solar radiation, cannot be realistically controlled in a field environment. Therefore, it is essential to manage those environmental factors such as water and nutrient availability to optimize phytoremediation potential. At this point, proper irrigation and fertilization have proven to be two necessary management techniques. Further research should test other management techniques, such as annual tillage or burning, for better turnover of vegetation during phytoremediation projects.

CHAPTER 4

Conclusions and Recommendations

This 2-year field demonstration quite clearly has shown that phytoremediation works. Degradation of total petroleum hydrocarbons (TPHs) was greater in all three vegetated treatments compared to the unvegetated control. The rate of remediation in all plots did not diminish with time; a "plateau" effect was not observed. During the remediation process, the hydrocarbons did not leach from the root zone, and the plants did not accumulate polycyclic aromatic hydrocarbons (PAHs) in the shoots.

The phytoremediation study at Craney Island was successful with statistical significance between all treatments at the end of the 2-year project. During the first year of the study, there was a trend showing an increase in contaminant reduction in the vegetated plots, but with no statistical significance. After the first growing season, contaminant reduction in the vegetated plots continued to increase rapidly. After 24 months of plant growth, TPH dissipation ranged from 40 to 50%. The unvegetated control plots had an average of 31% TPH dissipation by the end of the project. These results indicate that the presence of vegetation can accelerate remediation of aged hydrocarbons in soil. There was no evidence that TPH reduction had reached a plateau by the end of the 2-year field study. The presence of fescue enhanced the dissipation of PAHs relative to bioremediation in the absence of plants.

Laboratory and greenhouse studies were conducted in support of the field research. A shaking extraction method, quicker, less expensive, and with less variable results than soxhlet, was developed for analysis of aged petroleum contaminants in soil. A plant growth chamber study, designed to evaluate the fate of benzo[a]pyrene in the rhizosphere, indicated that benzo[a]pyrene is degraded in the rhizosphere and degradation by-products ultimately become incorporated into the soil matrix. Complete mineralization, volatilization, and plant uptake were minor. A greenhouse experiment, evaluating the effect of depth on phytoremediation, found that vegetated columns had lower residual concentrations of benzo[a]anthracene and benzo[a]pyrene in deeper soil layers than unvegetated columns. Phytoremediation may have a significant impact on contaminants in shallow subsoil.

The presence of vegetation offers many beneficial changes in the soil profile. The changes include chemical differences, biological changes, and physical properties. All of these interactive changes result in increased remediation potential as compared to unplanted contaminated soil. Chemical changes in the vegetated plots included not only the reduction of contaminant but also changes in pH and available nutrients. The unvegetated plots displayed a slightly higher pH when compared to the planted plots which may have been the result of unused nitrogen. Both grass species demonstrated an increased use of available nitrogen and more neutral pH as compared to the clover and bare plots. The rhizosphere soil of the vegetated plots supported greater microbial activity. There may have been several reasons for this increased activity, including exudates released from plant roots, easily biodegradable carbon source, and increased aeration within the root zone. Microbial numbers also were influenced by soil moisture. Plant mixtures may contribute significantly to optimal microbial activity.

One of the greatest differences between a vegetated soil profile and an unvegetated site is the physical change resulting from the presence of belowground biomass. Through the plant's extensive root system, a transport network is established throughout the soil. This network extends into the soil, providing a more uniform distribution of nutrients, water, and air in soil. Plant roots provide an ideal environment for degradation of organic compounds. Microorganisms associated with roots may be present in regions of the soil that might otherwise be inaccessible. A fine and fibrous root system does not appear to be critical for phytoremediation. Assessment of root establishment and turnover and associated phytoremediation efficiency should be included in future field studies.

The difference between soil in the vegetated and unvegetated plots was apparent in the field during sampling. The soil cores for the vegetated sites were well aggregated and unsaturated through the 24-in. (61-cm) soil profile. In the unvegetated plots, soil cores displayed poor aggregation and were generally saturated in the lower segments. Because phytoremediation is an aerobic process, the physical condition of the soil profile is very important. Both chemical and biological changes in the soil are closely tied to the physical condition.

There are several management practices that must be utilized and optimized for degradation of contaminants via phytoremediation. These include not only fertilization rates and irrigation schemes but also biomass removal and tillage. These practices have been shown to increase production in agricultural fields by returning easily degradable carbon to the field and enhancing aeration. Future studies should include development of management techniques.

To gain a better understanding of phytoremediation mechanisms, future research should include more intensive sampling of the physical characteristics in the soil profile. Monthly sampling for contaminant reduction analysis is too frequent due to relatively slow changes. Every 2 to 3 months would be adequate for tracking contaminant reduction. However, in order to monitor microbial activity, daily measurements of the soil moisture, soil air, and temperature profile should be taken. This would provide a continuous trend for analysis rather than monthly snapshots. Microbial activity could be closely monitored based on respiration and tied to the soil conditions. Contaminant degradation trends could then be linked to microbial activity and more accurately predicted.

Phytoremediation has broad applicability for contaminated sites that have growing seasons of reasonable length, a nonphytotoxic contaminant matrix, a hospitable climate, availability of appropriate seeds or plants, and contaminants that are accessible by roots. Implementation of phytoremediation is not technically complicated; however, expertise is needed for process optimization. Most phytoremediation projects require only a minimal amount of site management and can be implemented at a relatively low cost.

4.1 PHYTOREMEDIATION AS A VIABLE TECHNOLOGY

Although phytoremediation has a place in the remediation of petroleum-hydrocarbon-contaminated soils, there are limitations to the technology. Considerable time is needed to achieve regulated levels, depending upon the initial concentrations and the desired end point. The experience at Craney Island combined with other field tests suggests that roughly 50% of the petroleum hydrocarbons will be degraded in three growing seasons. For some applications, this rate will be too slow. Phytoremediation using the tested grasses and legumes is a reasonable alternative for surface contamination and will have minimal impact below 90 cm. Deeper rooted varieties of plants appear to be useful for a narrow range of organic contaminants.

4.2 MODES OF ACTION

The very detailed and broad analyses that accompanied this project have provided some insights into the mechanisms of phytoremediation. The entire soil biological system must be fully active before significant declines in TPH are observed. Active plant roots and microbes are crucial, and they require warm temperature, adequate soil moisture, and available nutrients such as nitrogen, phosphorus, and oxygen. In the cold winter months or during extended periods of drought or excessive moisture, the rate of remediation declines.

The plant roots appear to provide an ideal environment for degradation of organic compounds as a result of several mechanisms: (1) roots improve the structure of the soil, allowing rapid movement of water and gases through the soil; (2) microorganisms are able to accompany the roots into regions of the soil that might otherwise be inaccessible, particularly the interior of dense aggregates and similar microsites; and (3) the rhizosphere encourages high microbial populations and activities by improving transport of water and air, providing elevated concentrations of labile carbon through sloughing of cells and root exudation, and allowing rapid achievement of near-ideal moisture contents by encouraging water flow through the soil profile and removal of excess water after heavy rainfall. A very fine, fibrous root system is not critical for phytoremediation; white clover has a coarse structure, yet the plots with clover had some of the highest rates of TPH degradation.

Although much has been learned about degradation of petroleum hydrocarbons in the rhizosphere, many gaps in our knowledge still exist. The exact mechanism of degradation is not understood with confidence. It is not known if the plants are actively involved in the degradation process. The optimal concentrations of available nitrogen, phosphorus, and oxygen or moisture content are not known. The ideal root architecture is not certain. The entire remediation community is currently addressing the important question: Should total residual contaminants be measured or should only the bioavailable fractions be quantified?

4.3 FUTURE ROLES OF PHYTOREMEDIATION

This technology has broad applicability when the conditions are conducive to adequate plant growth and if the contaminants of interest are accessible to the roots. Growing seasons of reasonable length, a nonphytotoxic contaminant matrix, a hospitable climate, and availability of a source of seeds or plants are some of the critical elements. Although the implementation of phytoremediation is not technically complex, a certain level of expertise is needed to ensure that the entire system is designed to provide the best opportunity for success. Whether the technology is being used as a final polishing step or as the sole means of remediation, the minimal amount of site management and low cost will make it an attractive alternative.

Certain technical and practical deficiencies need to be addressed to aid in the entire remediation process. First, analytical methods need to be developed specifically for soils. The standard soxhlet extraction is inadequate under certain conditions. Second, current regulatory standards are not risk based. They are based on total extractable quantities of unknown materials. Regulations based upon potential toxicity or bioavailability are needed. Third, more information is needed about plant species that are best adapted to phytoremediation. Part of this information can come from simple screening, but a deeper understanding of the mechanisms involved is needed to enable the selection, breeding, or genetic engineering of plants specifically for phytoremediation.

CHAPTER 5

References for
Phytoremediation Demonstration

Al-Assi, A. 1993. Uptake of Polynuclear Aromatic Hydrocarbons by Alfalfa and Fescue, M.S. thesis, Department of Civil Engineering, Kansas State University, Manhattan.

Alef, K. and Nannipieri, P. 1995. *Methods in Applied Soil Microbiology and Biochemistry,* Academic Press, New York, 503–505.

Alexander, M. 1977. *Introduction to Soil Microbiology,* 2nd ed., John Wiley and Sons, New York, 423–437.

Alexander, M. 1982. Most probable number method for microbial population, in *Methods of Soil Analysis, Part 2, Chemical and Microbiological Properties,* Page, A.L., Miller, R.H., and Keeney, D.R., Eds., Soil Science Society of America, Madison, WI, 815–820.

Allison, L.E. 1960. Wet-combustion apparatus and procedure for organic and inorganic carbon in soil, *Soil Sci. Soc. Am. Proc.,* 24:36–40.

Anderson, T.A. and Walton, B.T. 1992. Comparative Plant Uptake and Microbial Degradation of Trichloroethylene in the Rhizospheres of Five Plant Species: Implications for Bioremediation of Contaminated Surface Soils, ORNL/TM-12017, Oak Ridge, TN, 186 pp.

Anderson, T.A., Guthrie, E.A., and Walton, B.T. 1993. Bioremediation in the rhizosphere — plant roots and associated microbes clean contaminated soil, *Environ. Sci. Technol.,* 27:2630–2636.

Aprill, W. and Sims, R.C. 1990. Evaluation of the use of prairie grasses for stimulating polycyclic aromatic hydrocarbon treatment in soil, *Chemosphere,* 20:253–266.

Atlas, R. and Bartha, R. 1992. Hydrocarbon biodegradation and oil spill bioremediation, in *Advances in Microbial Ecology,* Vol. 12, Marshall, K.C., Ed., Plenum Press, New York, 287–338.

Barber, S.A. 1984. *Soil Nutrient Bioaavailability,* John Wiley and Sons, New York.

Barber, D.A. and Martin, J.K. 1976. The release of organic substances by cereal roots into soil, *New Phytol.,* 76:69–80.

Bell, R.M. 1992. Higher Plant Accumulation of Organic Pollutants from Soils, EPA/600/SR-92/138, U.S. Environmental Protection Agency, Cincinnati, OH, 1–4.

Bell, R.M. and Failey, R.A. 1991. Plant uptake of organic pollutants, in *Organic Contaminants in the Environment,* Jones, K.C., Ed., Elsevier Science, New York, 189–206.

Blum, S.C. and Swarbrick, R.E. 1977. Hydroponic growth of crops in solutions saturated with [^{14}C] benzo[a]pyrene, *J. Agric. Food Chem.,* 25:1093–1096.

Borneff, J., Farkazdi, G., Glathe, H., and Kunte, H. 1973. The fate of polycyclic aromatic hydrocarbons in experiments using sewage sludge–garbage composts as fertilizers, *Zlb. Bakt. Hyg. Abt. I. Orig. B.,* 157:151–164.

Brilis, G.M. and Marsden, P.J. 1990. Comparative evaluation of soxhlet and sonication extraction in the determination of polynuclear aromatic hydrocarbons in soil, *Chemosphere,* 21:91–98.

Cerniglia, C.E. and Heitkamp, M.A. 1989. Microbial degradation of polycyclic aromatic hydrocarbons (PAH) in the aquatic environment, in *Metabolism of Polycyclic Aromatic Hydrocarbons in the Aquatic Environment,* Varanasi, U., Ed., CRC Press, Boca Raton, FL.

Chen, C.S., Rao, P.S.C., and Lee, L.S. 1996. Evaluation of extraction and detection methods for determining polynuclear aromatic hydrocarbons from coal tar contaminated soils, *Chemosphere,* 32:1123–1132.

Crowdy, S.H. and Jones, D.R. 1956. Partition of sulphonamides in plant roots: a factor in their translocation, *Nature,* 178:1165–1167.

Douglas, G.S., Bence, A.E., Prince, R.C., McMillen, S.J., and Butler, E.L. 1996. Environmental stability of selected petroleum hydrocarbon source and weathering ratios, *Environ. Sci. Technol.,* 30:2332–2339.

Douglas, G.S., McCarthy, K.J., Dahlen, D.T., Seavevy, J.A., Steinhauer, W.G., and Prince, R.C. 1992. The use of hydrocarbon analyses for environmental assessment and remediation, *J. Soil Contam.,* 1:197–216.

Douglas, G.S., Prince, R.C., Butler, E.L., and Steinhauer, W.G. 1994. The use of internal chemical indicator in petroleum and refined products to evaluate the extent of biodegradation, in *Hydrocarbon Bioremediation,* Hinchee, R.E., Alleman, B.C., Hoeppel, R.E., and Miller, R.N., Eds., Lewis Publishers, Boca Raton, FL.

Eckert-Tilotta, S.E. and Hawthorne, S.B. 1993. A supercritical fluid extraction with carbon dioxide for the determination of total petroleum hydrocarbons in soil, *Fuel A,* 72:1015–1023.

Edwards, N.T. 1983. Polycyclic aromatic hydrocarbons (PAHs) in the terrestrial environment — a review, *J. Environ. Qual.,* 12:427–441.

Edwards, N.T. 1986. Uptake, translocation, and metabolism of anthracene in bush bean (*Phaseolus vulgaris* L.), *Environ. Toxicol. Chem.,* 5:659–665.

Ferro, A.M., Sims, R.C., and Bugbee, B. 1994. Hycrest crested wheatgrass accelerates the degradation of pentachlorophenol in soil, *J. Environ. Qual.,* 23:272–279.

Finlayson, D.G. and MacCarthy, H.R. 1973. Pesticide residues in plants, in *Environmental Pollution by Pesticides,* Edwards, C.A., Ed., Plenum Press, New York.

Fisher, J.A., Scarlett, M.J., and Stott, A.D. 1997. Accelerated solvent extraction: an evaluation for screening of soils for selected U.S. EPA semivolatile organic priority pollutants, *Environ. Sci. Technol.,* 31:1120–1127.

Forrest, V., Cody, T., Caruso, J., and Warshawsky, D. 1989. Influence of the carcinogenic pollutant benzo[a]pyrene on plant development: fern gametophytes, *Chem. Biol. Inter.* 72:295–299.

Garland, J. and Mills, A. 1991. Classification and characterization of heterotrophic microbial communities on the basis of patterns of community-level sole-carbon-source utilization, *Appl. Environ. Microbiol.,* 57:2351–2359.

George, S. 1994. Bias associated with the use of EPA Method 418.1 for the determination of total petroleum hydrocarbons in soil, in *Hydrocarbon Contaminated Soils and Ground Water,* Vol. 4, Calabrese, E.J., Kostecki, P.T., and Bonazountas, M., Eds., Amherst Scientific Publishers, Amherst, MA, 115–142.

Gibson, D.T., Jerina, D.M., Yagi, H., and Yeh, J.C. 1975. Oxidation of the carcinogens benzo[a]pyrene and benzo[a]anthracene to dihydrodiols by a bacterium, *Science,* 189:295–297.

Graf, W. 1965. Uber naturliches Vorkommen und Bedeutung der kanzerogenen polyzyklischen, aromatischen Kohlenwasserstoffe, *Med. Klin.,* 60:561–565.

Gunther, T., Dornberger, U., and Fritsche, W. 1996. Effects of rye grass on biodegradation of hydrocarbons in soil, *Chemosphere,* 33:203–215.

Heath, J.S., Koblis, K., and Sager, S.L. 1993. Review of chemical, physical, and toxicologic properties of components of total petroleum hydrocarbons, *J. Soil Contam.,* 2:21.

Heitkamp, M.A. and Cerniglia, C.E. 1988. Microbial metabolism of polycyclic aromatic hydrocarbons: isolation and characterization of a pyrene-degrading bacterium, *Appl. Environ. Microbiol.,* 54:2549–2555.

Hsu, T.S. and Bartha, R. 1979. Accelerated mineralization of two organophosphate insecticides in the rhizosphere, *Appl. Environ. Microbiol.,* 37:36–41.

Jones, K.C. 1991. Contaminant trends in soil and crops, *Environ. Pollut.,* 69:311–325.

Kaplan, I., Lu, S.H., Lee, R.P., and Warrick, G. 1996. Polycyclic hydrocarbon biomarkers confirm selective incorporation of petroleum in soil and kangaroo rat liver samples near an oil well blowout site in the western San Joaquin Valley, California, *Environ. Toxicol. Chem.,* 15:696–707.

Khesina, A., Shcherback, N., Shabad, L.M., and Vostrov, M. 1969. Destruction of benzopyrene by soil microflora, *Byulleten Eksp. Biol. Med.,* 68:70–73.

Kirchmann, H. and Tengsved, A. 1991. Organic pollutants in sewage sludge: analysis of barley grains grown on sludge-fertilized soil, *Swed. J. Agric.,* 21:115–119.

Klevens, H.B. 1950. Solubilization of polycyclic hydrocarbons, *J. Phys. Colloid Chem.,* 54:283–298.

Knox, R.C., Sabatini, D.A., and Canter, L.W. 1993. *Subsurface Transport and Fate Processes,* Lewis Publishers, Boca Raton, FL.

Lappin, H.M., Greaves, M.P., and Slater, J.P. 1985. Degradation of the herbicide Mecoprop [2-(2-methyl-4-chlorophenoxy)propionic acid] by a synergistic microbial community, *Appl. Environ. Microbiol.,* 49:429–433.

Lee, E. and Banks, M.K. 1993. Bioremediation of petroleum contaminated soil using vegetation: a microbial study, *J. Environ. Sci. Health,* A28:2187–2198.

Mahro, B., Schaefer, G., and Kastner, M. 1994. Pathways of microbial degradation of polycyclic aromatic hydrocarbons in soil, in *Bioremediation of Chlorinated and Aromatic Hydrocarbons,* Hinchee, R.E., Leeson, A., Semprini, L., and Ong, S.K., Eds., Lewis Publishers, Boca Raton, FL.

Mueller, J.G., Chapmen, P.J., and Pritchard, P.H. 1989. Creosote contaminated sites: potential for bioremediation, *Environ. Sci. Technol.,* 23:1197–1201.

Mueller, J.G., Lantz, S.E., Blattmann, B.O., and Chapman, P.J. 1991. Bench-scale evaluation of alternative biological treatment processes for the remediation of pentachlorophenol- and creosote-contaminated materials: solid-phase bioremediation, *Environ. Sci. Technol.,* 25:1045–1055.

Ou, L.T., Jing, W., and Thomas, J.E. 1995. Biological and chemical degradation of ionic ethyllead compounds in soil, *Environ. Toxicol. Chem.,* 14:545–551.

Park, K.S., Sims, R.C., Dupont, R.R., Doucette, W.J., and Matthews, J.E. 1990. Fate of PAH compounds in two soil types: influence of volatilization, abiotic loss and biological activity, *Environ. Toxicol. Chem.,* 9:187–195.

Paul, E.A. and Clark, F.E. 1989. *Soil Microbiology and Biochemistry,* Academic Press, New York.

Pepper, I., Gerba, C., and Brendecke, J. 1995. *Environmental Microbiology,* Academic Press, New York.

Prince, R.C., Elmendorf, D.L., Lute, J.R., Hsu, C.S., Halth, C.E., Senlus, J.D., Dechert, G.J., Douglas, G.S., and Butler, E.L. 1994. 17a(H)21b(h)-Hopane as a conserved internal marker for estimating the biodegradation of crude oil, *Environ. Sci. Technol.,* 28:142–145.

Reddy, B.R. and Sethunanthan, N. 1983. Mineralization of parathion in the rice rhizosphere, *Appl. Environ. Microbiol.,* 45:826–829.

Reilley, K., Banks, M.K., and Schwab, A.P. 1996. Dissipation of polycyclic aromatic hydrocarbons in the rhizosphere, *J. Environ. Qual.,* 25:212–219.

Reimer, G. and Suarez, A. 1995. Comparison of supercritical fluid extraction and soxhlet extraction for analysis of native polycyclic aromatic hydrocarbons in soils, *J. Chromatogr. A,* 699:253–263.

Rovira, A.D. and Davey, C.B. 1974. *The Plant Root and Its Environment,* University Press, Charlottesville, VA.

Sandmann, E.R.I. and Loos, M.A. 1984. Enumeration of 2,4-D-degrading microorganisms in soils and the crop plant rhizosphere, *Chemosphere,* 13:1073–1084.

Sawyer, G.M. 1996. Determination of gasoline range, diesel range, and mineral oil range organics in soils and water by flame ionization gas chromatography, *J. Soil Contam.,* 5:261–300.

Schnitzer, M. 1982. Organic matter characterization, in *Methods of Soil Analysis, Part 2, Chemical and Microbiological Properties,* 2nd ed., Page, A.L., Miller, R.H., and Keeney, D.R., Eds., Soil Science Society of America, Madison, WI.

Schwab, A.P. and Banks, M.K. 1994. Biologically mediated dissipation of polyaromatic hydrocarbons in the root zone, in *Bioremediation through Rhizosphere Technology,* Anderson, T. and Coats, J., Eds., American Chemical Society Symposium Series 563, American Chemical Society, Washington, D.C., 132–141.

Shabad, L.M. and Cohan, Y.L. 1972. The contents of benzo[a]pyrene in some crops, *Gschwultsforsch.,* 40:237–243.

Sims, R.C. and Overcash, M.R. 1983. Fate of polynuclear aromatic compounds (PAHs) in soil–plant system, *Res. Rev.,* 88:1–68.

Stevenson, F.J. 1982. *Humus Chemistry: Genesis, Composition, Reactions,* John Wiley & Sons, New York.

Strand, S.E., Newman, L., Ruszaj, M., and Wilmoth, J. 1995. Removal of trichloroethylene from aquifers using trees, in *Proceedings of the National Conference on Innovative Technologies for Site Remediation and Hazardous Waste Management,* American Society of Civil Engineers, Pittsburgh, 605–612.

Thomey, N. and Sratberg, D. 1989. A comparison of methods for measuring total petroleum hydrocarbons in soil, in *Proceedings of the NWWA/API Conference on Petroleum Hydrocarbons and Organic Chemicals in Groundwater: Prevention, Detection, and Restoration,* National Water Well Association, Columbus, OH.

Veeh, R.H., Inskeep, W.P., and Camper, A.K. 1996. Soil depth and temperature effects on microbial degradation of 2,4-D, *J. Environ. Qual.,* 25:5–12.

Versar, Inc. 1995. Remedial Investigation/Feasibility Study, Fleet and Industrial Supply Center, Craney Island Fuel Terminal, Portsmouth, Virginia, Versar Environmental Risk Management, Inc., Springfield, VA.

Walter, U., Beyer, M., Klein, J., and Rehm, H.J. 1991. Degradation of pyrene by *Rhodococcus* sp. UW1, *Appl. Microbiol. Biotechnol.,* 34:671–676.

Walton, B.T. and Anderson, T.A. 1990. Microbial degradation of trichloroethylene in the rhizosphere: potential application to biological remediation of waste sites, *Appl. Environ. Microbiol.,* 56:1012–1016.

Walton, B.T. and Anderson, T.A. 1992. Plant–microbe treatment systems for toxic waste, *Curr. Opin. Biotechnol.,* 3:267–270.

Walton, B.T., Hoylman, A.M., Perez, M.M., Anderson, T.A., Johnson, T.R., Guthrie, E.A., and Christman, R.F. 1994. Rhizosphere microbial communities as a plant defense against toxic substances in soil, in *Bioremediation through Rhizophere Technology,* Anderson, T. and Coats, J., Eds., American Chemical Society Symposium Series 563, American Chemical Society, Washington, D.C., 82–92.

Weissenfels, W.D., Beyer, M., Klein, J., and Rehm, H.J. 1991. Microbial metabolism of fluoranthene: isolation and identification of ring fission products, *Appl. Microbiol. Biotechnol.,* 34:528–535.

Wild, S.R. and Jones, K.C. 1989. The effect of sludge treatment on the organic contaminant content of sewage sludges, *Chemosphere,* 19:1765–1777.

Wild, S.R. and Jones, K.C. 1992. Organic chemicals in the environment. Polynuclear aromatic hydrocarbon uptake by carrots grown in sludge-amended soil, *J. Environ. Qual.,* 21:217–225.

Wild, S.R., Jones, K.C., and Johnston, A.E. 1992. The polynuclear aromatic hydrocarbons (PAHs) content of herbage from a long-term grassland experiment, *Atmos. Environ.,* 26A:1299–1307.

Wollum, A. 1982. Cultural methods for soil microorganisms, in *Methods of Soil Analysis, Part 2, Chemical and Microbiological Properties,* 2nd ed., Page, A.L., Miller, R.H., and Keeney, D.R., Eds., Soil Science Society of America, Madison, WI, 781–814.

Yu, W., Dodds, W.K., Banks, M.K., Slaksky, J., and Strauss, E.A. 1995. Optimal staining and sample storage time for direct microscopic enumeration of total and active bacteria in soil with two fluorescent dyes, *Appl. Environ. Microbiol.,* 61:3367–3372.

PART **II**

Technology Design/Evaluation

J. Finn
Remediation Technologies, Inc.

Executive Summary

The Department of Defense (DOD) funded the Advanced Applied Technology Development Facility (AATDF) in 1993 with the mission of enhancing the development of innovative remedial technologies for the DOD by bridging the gap between academic research and proven technologies. AATDF is led by Rice University.

AATDF sponsored 12 projects that involve the quantitative demonstration of innovative remediation technologies. Field demonstrations have been completed for each of these projects. To further assist with the commercialization of the remediation technologies and to disseminate information on the technologies, AATDF prepared Technology Evaluation Reports (TERs) for many of the demonstrated technologies. These reports were designed to complement the project Technical Reports that were prepared by the principal investigator of each demonstration project. The TERs, which were prepared by an engineering firm independent of the principal investigators, focused on the design, cost and performance, and economics of each process or technology. The TERs also provided guidance for the applications of the technology.

This TER for phytoremediation was based primarily on results from the AATDF technology demonstration project entitled "Phytoremediation of Soil Contaminated with Hazardous Chemicals" (Banks et al., 1998), which was conducted on a field scale at the U.S. Navy's Craney Island Biotreatment Facility in Portsmouth, VA from September 1995 through October 1997.

The evaluation provided in this report was limited to the phytoremediation of aged hydrocarbons in soil. Other phytoremediation processes, including those addressing groundwater and extraction of metals or contaminants other than petroleum- and coal-derived hydrocarbons, were not addressed.

The AATDF phytoremediation project was performed by the following team:

- AATDF Project Manager: Stephanie Fiorenza
- Principal Investigators: M. Katherine Banks, A. Paul Schwab, and Rao S. Govindaraju, Purdue University (formerly of Kansas State University)
- TER Lead Author: John Finn, Remediation Technologies, Inc.

The overall objectives of the field test were to evaluate the use of vegetation to enhance bioremediation of soil contaminated with aged petroleum hydrocarbons on a field scale and to examine the effects of different plant types and rhizosphere characteristics on biodegradation of aged petroleum hydrocarbons.

The soils were characterized, and plant species were selected. Three separate planted treatments were established and monitored over time, with unvegetated plots serving as controls:

- Bermuda grass (*Cynodon dactyl* Vamont)/rye grass (*Lolium perene* Linn)

- Tall fescue (*Festuca arundinacea* Kentucky 31)
- White clover (*Trifolium repens* Dutch White)

Chemical analysis by gas chromatography of total petroleum hydrocarbons (TPHs) in the soil was the primary indicator used to assess phytoremediation. Leachate was collected and analyzed. Limited analyses of polycyclic aromatic hydrocarbons (PAHs) in soil and plant tissue were also conducted. In an attempt to remove heterogeneity from the soil PAH data, the concentrations were normalized to an oil-weight basis.

All of the plant species tested grew well in the petroleum-hydrocarbon-contaminated soil, which had an initial mean concentration of approximately 3000 mg/kg TPH. The greatest extent of TPH removal was measured in the fescue and clover plots. Over the 24-month period, TPH decreased by 45 to 50% in these plots. The decrease in the Bermuda grass plots was about 40%, and the bare, unvegetated control plots exhibited about a 30% decrease.

The trends observed in the TPH concentrations cannot be directly transferred to PAH concentrations. The greatest extent of PAH removal was observed in the fescue plots. However, the clover plots, which had high TPH removals, had one of the poorest performances for PAH removal. Over the 25-month period measured, PAH decreases ranged from 100% for fluorene to 25% for perylene in the fescue plots. The unvegetated control plots generally exhibited less removal, which ranged from 65% for fluorene to 26% for perylene (on an oil-weight basis).

The following conclusions were drawn from the results of the field test:

- The presence of plants resulted in a statistically significant enhancement of TPH and PAH degradation in soil when compared with unvegetated controls.
- The rate of degradation appeared to be dependent on the selection of plant species. Leguminous roots, root length, biomass, and surface area may be important parameters, but more research is needed to determine these relationships. Some of the results of this field test were suggestive of these mechanisms. For example, as described in the final report, in the clover treatments, degradation of TPH increased sharply during the period when the clover was dying and the roots were degrading. The hydrocarbons may have been degraded cometabolically, which would impact the more labile TPH compounds first and the recalcitrant PAHs last.
- Biodegradation by active microbial populations was indicated as the primary mechanism for decrease in TPH and PAH in this field test.
- The rate of degradation in all plots did not diminish with time. A plateau effect was not observed.
- The field test was one of the first examinations of the effect of plants on petroleum hydrocarbons done under controlled field conditions. The results of other similar studies are not yet available for comparison.

The design of future full-scale systems may differ substantially from the field test system. The field test was conducted on a relatively small scale and was primarily concerned with field-scale research to demonstrate the technology and to elucidate the factors that affect performance. The field test did not include a major design component because it used the existing prepared-bed biotreatment facility. The major design considerations for a full-scale system include soil characteristics, depth of treatment, area to be treated, nutrient amendments, toxicity of soil contaminants to plants, rainfall and climate, treatment end points, duration of treatment, and plant species. The full-scale design presented as an example does not include a prepared bed with a synthetic liner. Groundwater and leachate monitoring were used to address groundwater quality concerns. To optimize plant growth, most full-scale systems will require design of an appropriate irrigation system.

The analytical procedures used during the Craney Island phytoremediation field test were standard tests, with the exception of TPH measurements that were performed using a modified method. The standard EPA methods used for preparation and extraction of soil samples for the determination of TPH were modified to provide a methodology better suited to the particular needs of the project. Results of analyses performed by independent laboratories using EPA methods for TPH were subject to a high degree of variability and could not be correlated to the field study results. The variability was attributed to the testing of moist, unground samples.

The estimated cost for the field test at Craney Island totaled $320,000. This cost does not represent a reliable estimate for full-scale phytoremediation systems, because the field test was carried out primarily as a research project, and the small scale of the project tended to inflate the costs on a unit basis. The estimated unit cost was $240 per cubic yard or $160 per ton ($310 per cubic meter or $180 per metric ton).

The cost of a full-scale phytoremediation at Craney Island (the size of the entire biotreatment cell), treating 45,000 yd^3 (34,000 m^3) of soil, was estimated at $900,000 for 2 years of treatment with irrigation. This is equivalent to a unit cost of $20 per cubic yard or $13 per ton ($26 per cubic meter or $14 per metric ton). Economies of scale play an important role in decreasing the unit cost as the size of the project increases. These costs are similar to those of alternate treatments. The cost of solid-phase biological treatment (soil tilling) for an equivalent scale would be $24 per cubic yard or $16 per ton ($31 per cubic meter or $18 per metric ton). The slightly higher costs are attributable to the higher operating costs incurred by tilling.

Phytoremediation was shown in the Craney Island field test to result in only moderate decreases in hydrocarbon contaminant concentrations. The best treatment rate observed in the Craney Island field test, about 50% TPH dissipation over 24 months, was lower than the rate expected for more active solid-phase biological treatment by soil tilling. A typical solid-phase biological treatment system would result in at least 80% dissipation of moderately weathered diesel-range TPH over the same time period. Plant species and root development appeared to be significant factors leading to maximum treatment rates in the Craney Island field test. Other factors of importance were related to maintaining plant and microorganism growth.

The beneficial effects of phytoremediation may not only include decreased chemical concentrations, but may also include other forms of treatment such as decreased availability and toxicity and decreased potential for migration of contaminated soil particles from the site. The goal of these forms of treatment is to minimize the impacts to potential receptors in order to satisfy the short-term and long-term objectives of a site remediation plan. Ecological restoration may be an objective of a site remediation plan which could be addressed by phytoremediation.

As part of the economic analysis for this report, a cost model was developed to illustrate the relationships among the cost elements and to provide order-of-magnitude cost estimates. The major physical plant costs consist of irrigation equipment for locations requiring irrigation, on-site water production wells for sites needing a water source in addition to irrigation equipment, and synthetic liner installation for sites requiring prepared-bed systems. Operating costs include operational and environmental monitoring, fertilizing, harvesting, and project management and reporting.

The critical scale-up requirement and limitation for this technology are the need for more information regarding the mechanisms, performance, and economics of phytoremediation of hydrocarbon-contaminated soils. Larger sites will exhibit a greater variability of soil and contaminant characteristics. This variability will be an important consideration in determining the performance of larger systems. Other scale-up requirements and limitations that are of economic importance include water sources for irrigation and sources of large quantities of the specified plant seeds or cuttings.

Phytoremediation of hydrocarbons in soil is a slow-rate, low-cost technology that is anticipated to be most applicable to large volumes of soil containing low concentrations of contaminants treated at shallow depths. There are insufficient data regarding the performance of phytoremediation to define the full range of potential contaminants and concentrations which could be remediated by this technology. However, the results of recent studies, such as those reported at the IBC, Inc. Phytoremediation Conference, Seattle, WA, in June 1997 (Drake, 1997), suggest that phytoremediation may be applicable to a broad range of hydro-carbon mixtures, including crude oil, gas pit residues, refined petroleum products, and manufactured gas plant residuals in soil.

Phytoremediation of hydrocarbon-contaminated soil is potentially limited only by the ability to establish and maintain viable plant cover. The importance of species selection and consideration of native or locally adaptive species cannot be overemphasized. The results of the Craney Island field test suggest that criteria for plant selection will include plant main-tenance considerations in balance with degradation performance. The important site charac-teristics to be taken into account when considering this technology include temperature, latitude, precipitation, slope, soil characteristics, and depth of contamination. However, the relationship between these site characteristics and performance has not been quantitatively established.

Performance requirements that define the end points to be reached and compliance criteria which ensure safe and environmentally benign operations will have profound effects on the application of this phytoremediation technology. Regulatory performance requirements for sites with contaminated soils currently emphasize numerical cleanup standards. An emerging body of scientific evidence and site experience suggests that a regulatory framework for many sites in the future will be a risk-based approach to site management based on the actual availability and mobility of chemicals in the soil. Moreover, remedial performance objectives will more frequently include measures of ecological restoration. Phytoremediation has great potential to play an important role in this emerging regulatory context.

The operational compliance criterion that will have the most profound effect on the application of phytoremediation is the protection of groundwater during the process. It will be important to demonstrate that the risk posed, if any, by the hydrocarbons in the soil being treated is acceptable over the time period of the phytoremediation process and that phytoremediation will result in soil which no longer poses a threat to groundwater.

The remainder of this book is organized as follows:

- Chapter 6: "Summary of Technology Demonstration" — Summarizes the Craney Island phytoremediation field test.
- Chapter 7: "Engineering Design" — Describes the design aspects of full-scale phytoremediation of hydrocarbon in soil and provides an example of a full-scale design.
- Chapter 8: "Measurement Procedures" — Describes the measurement procedures used during the Craney Island field test and highlights measurement procedures recommended for a full-scale system.
- Chapter 9: "Costs and Economic Analysis" — Provides actual cost data for the Craney Island field test and estimated costs of full-scale systems. An economic analysis of the primary factors affecting cost is provided with a cost model.
- Chapter 10: "Performance and Potential Application" — Presents considerations affecting the application of this technology at sites in the U.S.
- Chapter 11: "References and Bibliography"

CHAPTER **6**

Summary of Technology Demonstration

6.1 INTRODUCTION

This Technology Evaluation Report was based primarily on results from the Advanced Applied Technology Development Facility (AATDF) technology demonstration project entitled "Phytoremediation of Soil Contaminated with Hazardous Chemicals." A summary of the demonstration project is provided in this chapter, including the technical approach, laboratory testing and initial studies, a description of the test site, field procedures, schedule, and summary of results. A detailed description of the AATDF demonstration is provided in Part I of this volume and in the Final Technical Report for the project (Banks et al., 1998).

6.2 SUMMARY OF THE TECHNICAL APPROACH

The overall objective of the field test was to evaluate the use of vegetation to enhance bioremediation of soil contaminated with aged petroleum compounds on a field scale at the U.S. Navy's Craney Island Biotreatment Facility. The subobjectives that guided the work were the determination of plant species needed to establish vegetation and enhance bioremediation in moderately contaminated soils and the definition of the rhizosphere characteristics that limit or enhance contaminant biodegradation.

The field test design consisted of the following four steps:

1. Characterize soil for chemical and physical properties and initial contaminant concentrations.
2. Select a variety of plant species suitable to the local environment.
3. Plant the selected species separately in contaminated soil located within a subarea of the biotreatment facility.
4. Monitor the progress of phytoremediation over time for each species and an unvegetated control. Chemical analysis of total recoverable hydrocarbons in the soil was the primary indicator used to assess phytoremediation.

6.3 SUMMARY OF LABORATORY TESTS AND GERMINATION STUDIES

An initial characterization of the physical and agronomic characteristics of the test area soil was conducted. Table 6.1 provides this characterization. Results indicated that the soil was a sandy loam with normal characteristics and was generally supportive of plant growth.

Table 6.1 Initial Soil Characterization for
Phytoremediation Field Test at
Craney Island Fuel Terminal,
Portsmouth, VA

	September 1995
pH	7.4
Electrical conductivity	4.0 mmhos/cm
NO_3-N	<0.1 mg/kg
NH_4-N	2.70 mg/kg
Bray-P	20.5 mg/kg
Cation exchange capacity	7.4 meq/100 g
Organic matter	4.4%
Ca	22.5 meq/l
Mg	14.8 meq/l
Na	12.0 meq/l
K	2.5 meq/l
Sodium adsorption ratio	2.8
Sand	60%
Silt	21%
Clay	19%
Texture	Sandy loam
Total organic carbon	1.8%
Solids	78.2%
Salt rank	Low

Seed germination studies were conducted using readily available seeds grown in the site test soil. Based on these studies and consultation with local horticultural and agronomic experts, the following plants were selected:

Common Name	Genus, Species, and Variety	Characteristics
Bermuda grass (sod) and rye grass (seed) (both species on the same plots)	*Cynodon dactyl* Vamont *Lolium perene* Linn	Warm-season perennial Cool-season annual
Tall fescue (seed)	*Festuca arundinacea* Kentucky 31	Rapidly growing cool-season grass, intensive root system
White clover (seed)	*Trifolium repens* Dutch White	Shallow-rooted legume

Unvegetated, untilled plots were used as controls. They were mixed when the soils were placed and fertilized.

6.4 TEST SITE DESCRIPTION

The test site was located at the Craney Island Fuel Terminal (CIFT) in Portsmouth, VA. The CIFT is the Navy's largest fuel facility in the U.S. The CIFT consists of over 1100 acres (445 ha) of both underground and aboveground fuel storage tanks. The phytoremediation field test was conducted in the Atlantic Division of Naval Facilities Engineering Command biological treatment cell located at the CIFT.

The bioremediation treatment cell is approximately 15 acres (6.1 ha) and is underlain by a compacted clay base, a 60-mil polyethylene liner, a synthetic geogrid, and approximately 12 in. (30 cm) of sand. The cell is also bermed on all sides, and sumps and pumps are used to collect and remove irrigation and storm water. Petroleum-contaminated soil is treated in the bioremediation cell by first placing a layer of soil in the cell and mixing with dedicated tilling and composting equipment. Tilling, irrigation, and fertilization then are performed to encourage bioremediation of the contaminants. A subarea of the bioremediation treatment cell containing a 2-ft (0.61-m) depth of soil was used for this phytoremediation demonstration project.

The demonstration field test site was a small area (100 ft [30.5 m] × 180 ft [55 m], approximately 0.5 acre (0.20 ha) located in the CIFT biotreatment cell. Intensive soil sampling after placement indicated a relatively homogeneous distribution of contaminants, which allowed a randomized block design. The study was divided into distinct plots of approximately the same size, as shown in Figure 6.1. Engineering drawings of the treatment area are provided in Chapter 7.

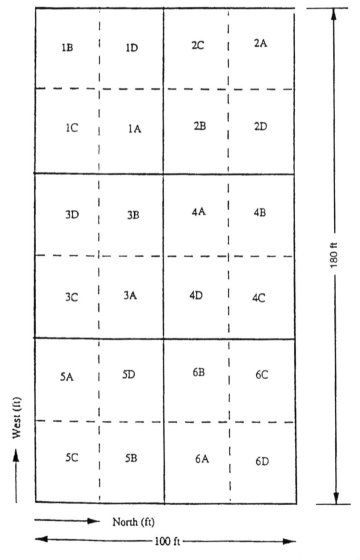

Figure 6.1 Schematic layout of the randomized block design. The number stands for the block number, and the letter stands for the treatment. (From Banks et al., 1996.)

6.5 PROCEDURES AND FIELD IMPLEMENTATION

This summary includes brief descriptions of the sampling procedures, establishment of vegetation, fertilization, irrigation, and key field observations:

- *Origin of the soil* — The soil used in the study originated in lagoons that were used at the CIFT from 1940 to 1978 for oil–water separation of ships' ballast and bilge waste. The lagoon sediments were excavated in 1995 and placed into the CIFT biotreatment cell using a front-end loader.
- *Soil sampling and analysis* — Soil sampling was conducted prior to seeding and then once per month or two during the test period, which was a 2-year period, from September 1995 through October 1997. The soils were analyzed for total petroleum hydrocarbons (TPHs) using a modified shaking extraction procedure followed by gas chromatography with flame ionization detection. Some soil samples were analyzed for polycyclic aromatic hydrocarbons (PAHs) using gas chromatography/mass spectrometry. Results are presented and discussed in Section 6.7.
- *Microbial activity* — The microbial activity of the soil samples was characterized during the course of the field test. An agronomic soil characterization was performed once per growing season.
- *Biomass* — Samples of aboveground plant mass were collected at three times during the first growing season and at the end of the second growing season. Samples of roots were collected once during each growing season. The mass was measured and several physical characteristics of the roots were measured. Some chemical analyses of plant material were conducted at the end of the field test.
- *Leachate* — Leachate from the test plots was collected by vacuum pore-water samplers. These samplers collected soil solution directly above the underlying sand layer. The samples were analyzed bimonthly, when leachate was available.
- *Tilling* — The soil in all plots had been disturbed during the placement of the soil onto the treatment bed. To create a seedbed, all plots were tilled once to a depth of approximately 0.5 ft (0.2 m) prior to the start of the field test.
- *Seeding and planting* — Vegetation was established by seeding with the typical amounts for each species. The Bermuda grass was established by laying sod.
- *Fertilization* — Fertilizer was applied initially during tilling and periodically during the field test. The rates of fertilization were chosen to provide a C:N:P ratio of 100:20:10 over the course of the field test period.
- *Irrigation* — The plots were irrigated as needed using a stationary spray system consisting of three stationary PVC pipe lines which remained in one position during the field test. Impact spray heads were mounted on riser pipes at spacings that provided even coverage. Applications of 2 to 3 in. (5.08 to 7.62 cm) of water per week were required during the summer months.

6.6 SCHEDULE SUMMARY

The 2-year field test was conducted according to the following schedule:

Task	Month/Year Completed
Work plan	July 1995
Initial soil characterization	July 1995
Vegetation plan (plant selection)	August 1995

Task	Month/Year Completed
Field test site established/time 0 sampling	September 1995
Phase I report	October 1995
Phase II report	March 1996
Phase III report	February 1997
Final sampling	October 1997
Final technical report	April 1998

6.7 SUMMARY OF RESULTS AND CONCLUSIONS

6.7.1 Field Results

The following key observations were made during the demonstration project:

- All of the plant species tested grew well in the petroleum-hydrocarbon-contaminated soil. The pattern of growth varied for each species. Bermuda grass was the first species to provide a full cover and establish roots. Bermuda grass was established by planting sod in the fall. Bermuda grass seed did not fare as well as sod during germination studies. Perennial rye grass seeded into these plots provided winter growth when the Bermuda grass was dormant. Fescue was established by seed and took longer to provide full cover than Bermuda grass, but grew well throughout the field test period and had the deepest root system. White clover was the slowest plant to become established, but provided a full cover by the end of the first growing season. Although white clover is a legume, few nitrogen-fixing root nodules were observed. The clover had the poorest growth of the species during the second, very dry, growing season, and weed species started to become established in the clover plots.
- Water management was an important parameter in maintaining plant viability throughout the growing season. Excess water in the winter and early spring of 1996 was slow in draining from the site. More importantly, a lack of irrigation in the summer of 1997, coupled with a drought during that period, resulted in low soil moisture content and deleterious effects on plant and microbial activity.
- The soil cores collected from the vegetated plots were well aggregated and well drained. However, cores from the unvegetated plots were observed to have poor aggregation and were generally saturated in the lower segments.
- Insect and disease damage was not significant.

6.7.2 Treatment Rates

The chemical analyses of the initial soil samples indicated that the hydrocarbon mixture of the test soil could be characterized as a weathered diesel product with an initial TPH concentration of about 3000 mg/kg. These chemical analyses were performed using an extraction procedure developed by the principal investigators (Banks et. al., 1998). Figure 6.2 presents chromatograms representing hydrocarbon mixture characteristics before and after the first growing season.

Table 6.2 and Figure 6.3 present the summary of TPH results. The greatest extent of TPH removal was measured in the clover plots. Over the 24-month period, TPH decreased by about 50% in these plots. The decrease in the fescue plots was about 45%, and in the Bermuda grass plots, the decrease was about 40%. The bare, unvegetated control plots exhibited a 31% decrease. Thus, the vegetated treatments exhibited a greater extent of degradation, from 9 to

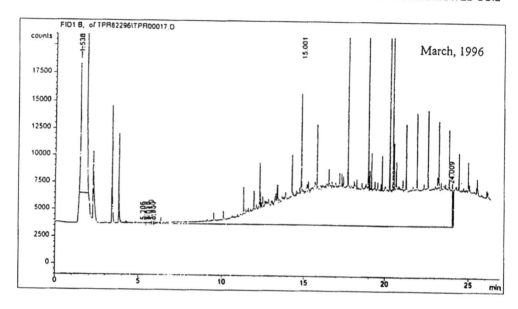

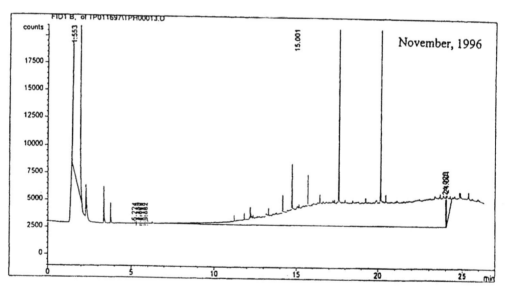

Figure 6.2 Representative gas chromatograms of hydrocarbons in test soil. (From M.K. Banks et al., 1996.)

19% more, than the bare controls. A statistical analysis was performed by the principal investigators on the percent change in TPH concentration. The analysis of variance yielded the mean separation test used to determine the statistical differences among the means at each sampling time. Additional information regarding the statistical methods and results is presented in the final report (Banks et al., 1998).

Similarly, Table 6.3 presents the summary of PAH results. In an attempt to remove heterogeneity from the PAH concentration data, the concentrations were normalized on an extract weight basis by A.D. Little. To accomplish this, the gravimetrically determined extract weight (post alumina cleanup) was substituted for the soil weight in the concentration calculation. The concentrations were thus the concentration of the particular PAH in the

Table 6.2 Loss of TPH in Soil During Field Test of Phytoremediation of Petroleum-Hydrocarbon-Contaminated Soil, Craney Island, Portsmouth, VA[a]

Treatment	November 1995	March 1996	July 1996	November 1996	March 1997	July 1997	October 1997
Clover	0%a	8%a	15%a	29%ab	30%a	34%(ab)	50%a
Fescue	0%a	5%a	12%a	33%a	31%a	42%(a)	45%b
Bermuda	0%a	4%a	6%a	27%ab	27%a	36%(ab)	40%c
Unvegetated	0%a	2%a	9%a	21%b	21%a	29%(b)	31%d

[a] Suffixes a, b, c, and d indicate statistically equivalent values ($p = 0.05$ or [0.1]) among treatments on each sampling date. For example, for the November 1996 data, the results for clover, fescue, and Bermuda grass are all statistically equivalent, while the results for clover, Bermuda, and unvegetated are statistically equivalent. By the end of the study, in October 1997, each treatment had a statistically different result.

extracted oil, not the soil. Not only did this reduce the effects of heterogeneous TPH concentrations, but it also allowed evaluation of the degradation of the actual oil components.

The trends in the PAH data were very consistent. For nearly every compound, the percent degradation was the greatest in the fescue plots and the least in the unvegetated plots. Very often, the percent degradation in the clover plots was statistically less than in the fescue and equivalent to that in the unvegetated plots.

The trends observed in the TPH concentrations could not be directly transferred to PAH concentrations. The greatest extent of PAH removal was measured in the fescue plots. However, the clover plots, which had high TPH removals, had one of the poorest performances for PAH removal. Over the 25-month period, PAH decreases ranged from 100% for fluorene to 25% for perylene in the fescue plots. The unvegetated control plots exhibited less removal. The decrease in PAHs ranged from 65% for fluorene to 26% perylene (on an oil-weight basis).

Figure 6.4 shows the relationship of root biomass and root length for the three test species. Clover had substantially less root mass than fescue or Bermuda grass.

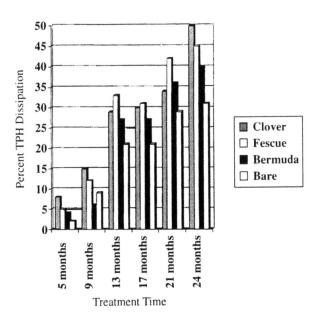

Figure 6.3 TPH dissipation: loss of TPH in soil during field test of phytoremediation of petroleum-hydrocarbon-contaminated soil, Craney Island, Portsmouth, VA.

Table 6.3 PAH Degradation Using Oil-Based Weight Method by A.D. Little

Compound	Number of Rings	% Degradation				LSD[a] 0.05 (0.1)
		Clover	Unvegetated	Fescue	Bermuda	
C1-Naphthalene	2	1.4	30	29	−15	ns
C2-Naphthalene	2	53ab	58a	49ab	8b	48
C3-Naphthalene	2	80a	73ab	71ab	49b	27
C4-Naphthalene	2	87a	86a	71ab	61b	19
Acenaphthylene	3	−60	−22	36	7	—
Acenaphthene	3	66	83	80	68	ns
Biphenyl	2	−24b	15ab	22ab	37a	57
Fluorene	3	3	62	75	75	62 ns
C1-Fluorene	3	58b	65b	100a	84ab	30
C2-Fluorene	3	84ab	68b	100a	87ab	32
C3-Fluorene	3	91ab	71ab	100a	85ab	28
Anthracene	3	32c	42bc	65a	53ab	17
Phenanthrene	3	58	63	75	54	ns
C1-Phenanthrene/anthracene	3	50(ab)	41(b)	80(a)	63(ab)	ns (33)
C2-Phenanthrene/anthracene	3	74	62	86	70	ns
C3-Phenanthrene/anthracene	3	82(ab)	68(b)	86(a)	76(ab)	ns(16)
C4-Phenanthrene/anthracene	3	85a	74ab	81a	63b	14
Dibenzothiophene	3	46	59	72	53	ns
C1-Dibenzothiophenes	3	77	60	80	57	ns
C2-Dibenzothiophenes	3	85(ab)	65(b)	88(a)	73(ab)	ns (20)
C3-Dibenzothiophenes	3	88a	66b	89a	79ab	18
Fluoranthene	4	79b	80b	93a	80b	6
Pyrene	4	82ab	72b	93a	75ab	14
C1-Fluoranthene/pyrene	4	73ab	62b	84a	70ab	16
C2-Fluoranthene/pyrene	4	58ab	52b	69a	56ab	15
C3-Fluoranthene/pyrene	4	50ab	36c	59a	45bc	13
Benzo[a]anthracene	4	71b	61b	86a	64b	13
Chrysene	4	60b	55b	78a	54b	13
C1-Chrysene	4	59bc	53 c	74a	68ab	15
C2-Chrysene	4	54ab	44b	65a	60a	14
C3-Chrysene	4	23b	28b	50a	49a	20
C4-Chrysene	4	25	26	37	27	ns
Benzo[b]fluoranthene	5	43b	59ab	67a	49b	18
Benzo[k]fluoranthene	5	30(b)	48(ab)	53(a)	45(ab)	ns (20)
Benzo[e]pyrene	5	32	30	37	27	ns
Benzo[a]pyrene	5	49b	51b	69a	53b	15
Perylene	5	−8b	26a	25a	−3b	19
Indeno[1,2,3-c,d]pyrene	6	−42	14	30	22	—
Dibenzo[a,h]anthracene	5	37	35	43	45	ns
Benzo[g,h,i]perylene	6	−35	19	35	22	—

[a] LSD = least significant difference, 0.05 = using a 95% confidence interval and (0.1) = using a 90% confidence interval.

Microbial plate counts and microbial activity, as measured using the BIOLOG culturing system with multiple substrates, indicated active microbial populations in all plots. Differences in the characteristics of microbial populations among the treatment plots were observed during the first year. Petroleum degradation was significantly higher in the clover plots.

Some of the results of this field test suggested why the decreases observed in TPH concentration could not be directly transferred to decreases in PAH concentration. For example, as described in Part I, in the clover treatments, degradation of TPH increased sharply during the period when the clover was dying and the roots were degrading. The hydrocarbons possibly were degraded cometabolically, which would impact more labile TPH compounds first and the recalcitrant PAHs last.

The following conclusions are supported by the results of the field demonstration:

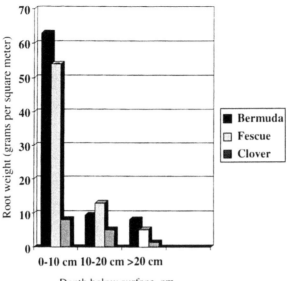

Figure 6.4   Root biomass (1 year) field test of phytoremediation of petroleum-hydrocarbon-contaminated soil, Craney Island, Portsmouth, VA.

- The presence of plants resulted in a statistically significant enhancement of dissipation of TPH and PAH in the soil when compared with unvegetated controls.
- The rate of degradation appeared to be dependent on the selection of plant species. By the end of the 2-year test period, the clover, Bermuda/rye grass, and fescue each had statistically different TPH degradation results.
- The extent of TPH degradation was not simply directly proportional to plant growth or root density. The clover species exhibited the greatest TPH degradation and yet had the lowest root and shoot biomass. Its roots were more coarse and less deep than the other species. Its sustained growth would have required more irrigation than the other species. Thus, the species providing the highest degradation rate may not be the species with the highest growth rates or the species that performs best under low-maintenance conditions.
- Biodegradation by active microbial populations was indicated as the primary mechanism for decrease in TPH and PAH in this field test. The presence of vegetation most likely affected degradation indirectly by providing an enhanced chemical, biological, and physical environment which promotes microbial action.
- No evidence of significant plant uptake of hydrocarbon contaminants was found. Therefore, phytoremediation of petroleum-contaminated soils should pose no threat to animals that graze on the plants.
- The rate of degradation in all plots did not diminish during the 2-year period of the field demonstration. A plateau effect was not observed, suggesting that long-term monitoring of the site would show greater decreases in TPH and PAH concentration than those observed at the end of 2 years.
- This field test was one of the first examinations of the effect of plants on petroleum hydrocarbon done under controlled field conditions. The results of other similar studies are not yet available for comparison.

CHAPTER 7

Engineering Design

This chapter describes the engineering design aspects of the type of phytoremediation system used in the Craney Island field test. An example of a full-scale design is presented. Additional full-scale design considerations are discussed in Sections 9.3 and 10.2.

7.1 CLASSIFICATION OF TREATMENT SYSTEMS USED IN PHYTOREMEDIATION OF HYDROCARBONS IN SOIL

Tables 7.1 and 7.2 identify the primary and secondary treatment systems to be generally included in the phytoremediation of hydrocarbons in soil, in accordance with the standard classifications of the "Guide to Documenting Cost and Performance for Remediation Projects" (Federal Remediation Technologies Roundtable [FRTR], 1995). The Craney Island field test used soil that was excavated and placed into a treatment area. Therefore, the primary treatment system was considered to be an *ex situ* solid-phase bioremediation. However, aside from the initial disturbance of soil (which may have had significant effects), the field test was similar to an *in situ* process. Supplemental treatment systems included mixing, screening, and shredding of the soil prior to placement. The subsequent phytoremediation used nutrient additions to maintain plant and microbial growth. *In situ* bioremediation is also a viable phytoremediation treatment option for surface soils.

7.2 PROCESS FLOW DIAGRAM

Phytoremediation is more akin to an agricultural activity than an engineered process. In contrast to other Advanced Applied Technology Development Facility demonstration technologies, it is not conducive to analysis and modification using conventional chemical engineering process flow diagrams and piping and instrumentation diagrams. Mass and energy balances around phytoremediation are difficult to obtain due to the open and dynamic nature of the process. Nonetheless, this section provides a process flow schematic and conceptual mass balance. No direct energy inputs, aside from sunlight, were required, and therefore an energy balance is not included in this report. Indirect energy inputs include pumped water and tractor/harvest power.

Figure 7.1 presents the schematic flow diagram of phytoremediation of hydrocarbons in soil. Inputs of soil, water, fertilizer, carbon dioxide, oxygen, sunlight, and seed are listed on the left side of the diagram, and outputs of evapotranspiration, runoff, leachate, plant biomass, and soil humic material are listed on the right. The soil in itself represents a complex

Table 7.1 Primary Treatment Systems[a] Used in the Phytoremediation Demonstration

Soil *In Situ*	Soil *Ex Situ*	Groundwater *In Situ*	Groundwater *Ex Situ*
Not applicable	Solid-phase bioremediation	Not applicable	Not applicable

[a] From FRTR (1995) Tables 4 and 5. Derived from EPA's VISITT database and a screening matrix prepared jointly by EPA and Air Force personnel.

mixture of inorganic minerals, organic colloidal material, hydrocarbon contaminants, micro-organisms, and invertebrate animals.

The inputs result in the germination and growth of plants that fix carbon from the atmosphere and produce biomass in the form of roots and aboveground shoots. Some of the biomass is removed from the system by harvesting, while the remainder is incorporated into the soil as humic material. The roots penetrate the soil, providing oxygen, carbon sources, nutrients, and enzymes, all which stimulate the growth of microorganisms in the root zone. The increased number of microorganisms can promote biodegradation. A series of harvest and regrowth cycles occurs during the course of the growing seasons. The resulting soil has lower total concentrations of petroleum hydrocarbons and increased humic material due to the increased numbers and activity of rhizosphere microorganisms.

Evapotranspiration is the total water lost by surface evaporation and transpiration of water by plant tissue. In a system using a synthetic liner, runoff and leachate are collected and either evaporated or recycled as irrigation water. The synthetic liner prevents losses of leachate to groundwater. A system that does not have a synthetic liner would use leachate and ground-water monitoring to address groundwater quality concerns.

The primary concerns of a mass balance on the phytoremediation system are in regard to possible contaminant losses from the system, which could impact potential receptors, and to document the fate of petroleum hydrocarbons in the system. In the Craney Island field test, losses of hydrocarbons in the leachate were below detection limits of 1 mg/l total petroleum hydrocarbon (TPH). This result corresponds to leaching from the soil of less than 1 mg/kg during a season of rainfall minus evapotranspiration. Losses due to volatilization were not measured but were estimated to be negligible because the hydrocarbon at the site was an aged diesel. Losses of unmetabolized petroleum hydrocarbons in the aboveground plant tissue were below detection limits. Thus, the mass balance with respect to hydrocarbon contaminants is reduced to the simple percentage of TPH and polycyclic aromatic hydrocarbon dissipation presented in Tables 6.2 and 6.3.

A potentially important remedial effect which is not revealed by the mass balance is the degree to which the hydrocarbon contaminants may have become more tightly bound within the soil matrix, and therefore less available to potential receptors, as a result of the phytoremediation process. This aspect of the process is further discussed in Section 10.5.

Table 7.2 Supplemental Treatment Systems[a] Used in the Phytoremediation Demonstration

Pretreatment (Solids)	Augmentation (for *In Situ* Process)	Posttreatment (Air)	Posttreatment (Solids)	Posttreatment (Water)
Mixing Nutrient injection Screening Shredding	Not applicable	Not applicable	Not applicable	Not applicable

[a] From FRTR (1995) Tables 4 and 5. Derived from EPA's VISITT database and a screening matrix prepared jointly by EPA and Air Force personnel.

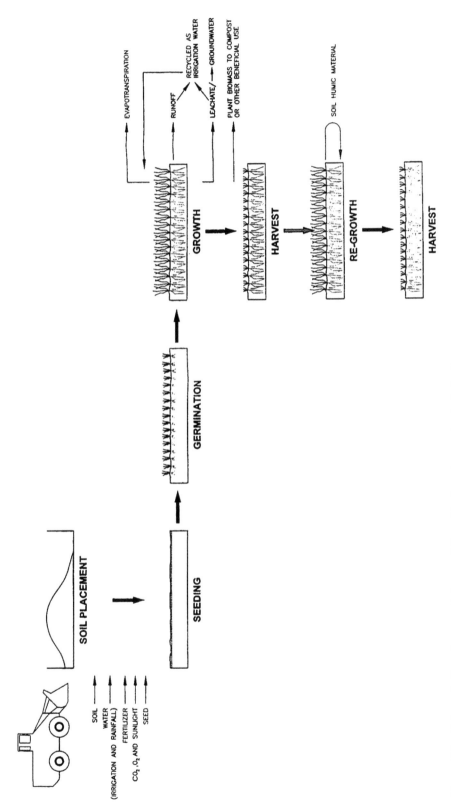

Figure 7.1   Schematic flow diagram: phytoremediation of hydrocarbon-contaminated soils.

7.3 PROCESS DESIGN AND ENGINEERING REQUIREMENTS

The design of phytoremediation systems for treatment of hydrocarbons in soil will include several process design considerations and associated constraints. For example, the following considerations and constraints were important in the design of the Craney Island field test:

Design Consideration	Site-Specific Constraints
Origin of the soil	The soil used in the study originated in lagoons that were used at the Craney Island Fuel Terminal (CIFT) from 1940 to 1978 for gravity oil–water separation of ships' ballast and bilge waste. The lagoon sediments were excavated in 1995 and placed into the CIFT biotreatment cell using a front-end loader.
Soil characteristics	The soil was provided as excavated from the subsurface at the CIFT.
Depth of treatment	The soil depth was as placed in the existing biotreament cell to a depth of 2 ft (0.6 m).
Area to be treated	The area was constrained by what was considered to be readily watered and harvested, and by the area limitations imposed by the ongoing biotreatment operation in the remainder of the biotreatment cell.
Nutrients	Nutrients (C:N:P) were added in the standard ratio of 100:20:10.
Toxicity	Plants were selected that could tolerate the soil conditions at the site. In selecting sites for future field demonstrations of hydrocarbon phyto-remediation, sites may be selected which do not contain toxic concentrations of heavy metals or other constituents which could affect plant growth and confound the evaluation of performance.
Rainfall/climate	The site represented temperate, nonarid Eastern Seaboard climatic conditions. Irrigation was a design consideration because of the possibility of drought periods.
Treatment end points	The only numerical treatment end point was provided by the state of Virginia as 750 mg/kg TPH. This end point changed to a risk-based approach adopted by the state during the project.
Treatment duration	The project had a duration of an initial partial growing season (September through November) followed by two full growing seasons. Duration was constrained by project contract considerations and local site operational considerations.
Plant species	Plant species were selected to be compatible with the local climate and soil type and to represent cool- and warm-climate grasses and a species of legume.

A plan view of the existing biotreatment facility and the phytoremediation subarea is shown in Figure 7.2. A schematic cross-section of the prepared-bed biotreatment cell used for the field test is shown in Figure 7.3. The cell was underlain with a geogrid and an impermeable synthetic liner, over which a layer of sand was placed as a drainage medium. The depth of the soil in the phytoremediation plots was 2 ft (0.6 m).

7.4 FULL-SCALE DESIGN

The Craney Island site was used as a basis for design of a full-scale phytoremediation system. The system was designed to provide phytoremediation of 45,000 yd^3 (34,405 m^3) of hydrocarbon-contaminated soil. The soil would be moved and spread 2 ft (0.6 m) deep across a 1000-ft × 600-ft (305-m × 183-m) fenced area.

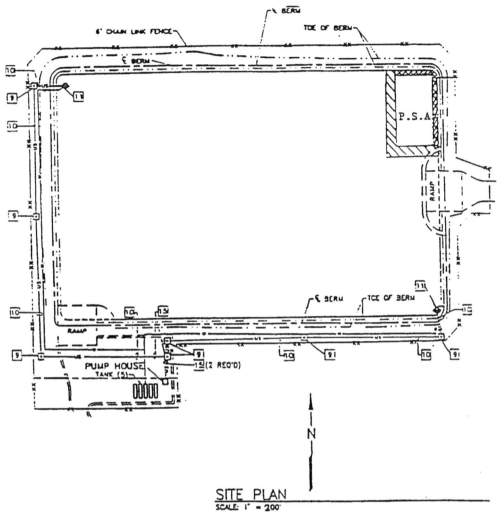

Figure 7.2   Biological treatment facility. P.S.A. = phytoremediation study area. Approximately 180 ft × 100 ft. Exposed buffer zone = 20 ft (▨). Edge buffer zone = 10 ft (▨).

It was assumed that the system would not require a prepared bed with a synthetic liner. Monitoring wells would be installed upgradient and downgradient of the system for ground-water monitoring. Pipe lysimeters would be installed at four locations within the treatment bed to allow monitoring of leachate quality. The system would be seeded and irrigated and then periodically harvested over several growing seasons.

Design drawings and calculations are provided in Appendix 2.

7.5 PROCESS EQUIPMENT

This section makes recommendations for full-scale construction and operation equipment. The descriptions of process equipment are presented in the order in which they appear in the process flow diagram (Figure 7.1).

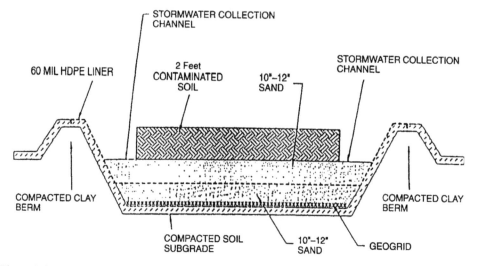

Figure 7.3   Schematic cross-section of the biological treatment cell Fleet and Industrial Supply Center, Craney Island Fuel Terminal, Portsmouth, VA. (From Banks et al., 1996.)

7.5.1 Soil Placement

No special equipment is required for soil placement. However, it is important to avoid excess compaction of soils by avoiding heavy equipment travel on the placed soil. Conventional front-end loaders and bulldozers may be used to place and spread the soil. Screening and shredding of some of the soils may occur immediately prior to placement using conventional soil classification equipment such as grizzly bar screens and power shaker screens.

Large systems involving the hauling and placement of more than 10,000 yd^3 (7600 m^3) may benefit from the use of self-loading, bottom-dumping scrapers to efficiently haul and spread the soil.

7.5.2 Seeding/Sod Placement

Full-scale seeding will be most efficiently done using hydroseeding equipment. Commercial hydroseeders spread an even mixture of seed, fertilizer, tackifier, and mulch.

7.5.3 Fertilizer/Lime

Full-scale systems can use agricultural tractor-mounted implements to apply fertilizer and lime.

7.5.4 Irrigation

Full-scale systems will use standard agricultural irrigation systems. Solid-set systems offer the lowest operation and maintenance. Other types of irrigation include move-stop and continuous-move systems (U.S. EPA, 1981). The system and sizing will depend on the labor available at the site and the application rate required for the climate and type of vegetative cover.

7.5.5 Drainage

Most full-scale systems treating hydrocarbons in soil will not have a synthetic liner. Drainage considerations are a concern in these systems if the soil infiltration rate is lower than common precipitation and snowmelt events. In these instances, the field should be sloped to a low point where runoff storage is provided.

7.5.6 Harvesting

Full-scale systems will use standard agricultural hay cutting and bailing equipment as needed during the growing season.

CHAPTER 8

Measurement Procedures

This chapter lists the measurement procedures recommended for full-scale phytoremediation of hydrocarbons in soil. Site characteristics and operating parameters are presented in Table 8.1, in accordance with the Federal Remediation Technologies Roundtable (FRTR) format. Measurement procedures for operating parameters that have potential effects on treatment cost or performance are presented in Table 8.2, in accordance with the FRTR format.

Among the most critical measurement procedures are hydrocarbon analyses. A time-series set of soil samples is analyzed to determine percentage of hydrocarbon removed. The extraction of hydrocarbons from the soil matrix is a critical step in these analyses. The efficiency of extraction can be impacted by the moisture content and soil chemistry, and great care should be taken in designing the analytical program.

A shaking extraction procedure was developed for the phytoremediation field test as an alternative to soxhlet extraction. The shaking extraction method was found to be equivalent to the soxhlet method and was quicker and used less solvent (Schwab et al., submitted).

As described in Section 10 for risk-based cleanup actions, measurement of bioavailability, mobility, and toxicity becomes essential. Specific procedures for these measurements should be tailored to the regulatory requirements and site conditions.

Table 8.1 Parameters Affecting Cost and Performance in Phytoremediation Applications

Site Characteristics	Operating Parameters
Existing vegetation	Moisture content
Soil classification (texture)	pH
Hydraulic capacity	Residence time
Moisture content	Temperature
Water retention	Rainfall
pH	Irrigation rate
Salinity	Microbial concentration
Cation exchange capacity (CEC)	Microbial activity: oxygen uptake rate, carbon dioxide evolution, BIOLOG
Buffer capacity (lime requirement)	Root and shoot mass concentrations
Sodium hazard (of irrigation water)	Ecological restoration goals
Inorganics in soil: chlorides, sulfate, nitrate, phosphate, ammonia	Nutrients and other soil amendments
Total organic carbon (TOC)	Total hydrocarbon concentrations in soil, leachate, and plant biomass
Total petroleum hydrocarbons (TPHs)	Leachable hydrocarbon concentrations
Benzene, toluene, ethylbenzene, and xylenes (BTEX)	Toxicity
Total metals, such as arsenic, barium, boron, cadmium, chromium, copper, lead, mercury, molybdenum, nickel, selenium, vanadium, and zinc	
Leachable concentrations of chemical constituents	

Table 8.2 Site Characteristics: Measurement Procedures and Potential Effects on Treatment Cost or Performance

Matrix Characteristics	Measurement Procedures[a]	Important to Document Procedure?	Potential Effects on Cost or Performance
Existing vegetation	Surveys of existing vegetation at the site locale document the species, growth characteristics, and prevalence.	Yes	A predesign survey of existing vegetation is critical to selecting species to use in the phytoremediation system which will most likely thrive in the given climate and soil conditions with the least input.
Soil classification (texture)	Soil classification is a semiempirical measurement of sand, silt, clay, gravel, and loam content. Several soil classification schemes are in use and include the ASTM Standard D 2488-90 and the USDA and CSSC systems.	Yes	Soil classification is an important characteristic for assessing the effect on cost or performance of phyto-remediation soil characteristics critical for plant selection.
Hydraulic capacity	Hydraulic capacity of surface soil is measured by standard basin infiltration tests. Results are typically reported in units of inches of water per hour.	Yes	This characteristic is important in designing water management of the phytoremediation system. For example, irrigation systems should not exceed the hydraulic capacity.
Moisture content	Soil moisture content is typically measured using a gravimetric ASTM standard, D2216-90, Test Method for Laboratory Determination of Water (Moisture) Content of Soil and Rock, also ASA No. 9 21-2.2. The use of tensiometers to measure field capacity has been problematic in soils contaminated with hydrophobic compounds.	No	The moisture available to plants is a critical matrix characteristic for plant selection.
Water retention	Standard Method ASA No. 9 26-6	No	Water retention is an important soil characteristic for plant selection and irrigation design.
pH	EPA SW-846 Method 9045 and ASTM methods for soil (ASTM D 4972-89, Test Method for pH of Soils) and groundwater (ASTM D 1293-84).	No	The pH of the matrix can impact the solubility of contaminants and all biological activity and can also affect the operation of phytoremediation.
Salinity	Electrical conductivity (EC) by Standard Agronomy Method ASA No. 9 10-3.3.	No	High salinity, as measured by EC, is toxic to many plants and must be taken into account in plant selection.
CEC	Standard Agronomy Method ASA No. 9 8-3. It is a rough index of all reactions with charged colloidal surfaces.	No	Soils with a high CEC have high capacity for retaining cations.
Buffer capacity (lime requirement)	Standard Agronomy Method ASA No. 9 12 3.4.5	No	Important for determining the lime requirement of overdue soils.

Parameter	Method		Comments
Sodium hazard	Sodium adsorption ratio USDA 60 using measurements of sodium, calcium, and magnesium ions in soil solutions.	No	Soils and irrigation waters with high sodium content result in soils that crust, swell, or disperse, decreasing permeability and adversely affecting plant growth.
Inorganics in soil: chlorides, sulfate, nitrate, nitrite, phosphate, ammonia	Standard Agronomy Methods, ASRM series, or EPA Method 300.0 modified	Yes	These measures are critical to assessing soil fertility and potential chloride toxicity.
TOC	EPA SW-846 Method 906, Agronomy Method ASA No. 9 33-6	No	TOC affects the desorption of contaminants from soil and thus impacts the availability of contaminants to plants and microorganisms and the extent of removal possible.
TPH	EPA SW-846 Method 8015. The recommended TPH measurement by the TPH Working Group Methodology does not include nonpetroleum fractions, such as animal fats and humic and fulvic acids.	Yes	TPH is often a key measure of hydrocarbon contamination used for regulatory compliance. TPH affects the desorption of contaminants from soil. Elevated levels of TPH may result in agglomeration of soil particles.
Polycyclic aromatic hydrocarbons (PAHs)	EPA SW-846 Method 8270B can be used for high concentrations; lower concentrations may require GC/MS with SIM.	Yes	PAHs are often key measures for regulatory compliance.
BTEX and other volatile organics	EPA SW-846 Method 8260A.	Yes	May be important for regulatory compliance.
Total metals, such as arsenic, barium, boron, cadmium, chromium, copper, lead, molybdenum, mercury, nickel, selenium, vanadium, and zinc	EPA SW-846 Method 6010A and 7471 for mercury.	Yes	May be important for regulatory compliance and for assessment of potential toxicity to plants.

[a] ASA = American Society of Agronomy, ASTM = American Society for Testing and Materials, EPA = U.S. Environmental Protection Agency, USDA = U.S. Department of Agriculture.

CHAPTER 9

Costs and Economic Analysis

This chapter reports the costs of the Craney Island phytoremediation technology demonstration and provides estimated costs for full-scale phytoremediation. A discussion of factors affecting the cost and performance of full-scale systems is included.

9.1 ESTIMATED COST OF FIELD DEMONSTRATION

The estimated cost for conducting a field test similar to that conducted at Craney Island totals $320,000. With a soil volume of 100 ft × 180 ft × 2 ft (30 m × 50 m × 0.6 m) (1333 yd^3) (1021 m^3), the resulting unit costs would be $ 240 per cubic yard ($183 per cubic meter). These costs do not represent reliable estimates of actual phytoremediation systems, because the work was carried out primarily as a research project and because the small scale of the project tended to inflate the unit costs. Cost estimates for full-scale phytoremediation systems are discussed in Section 9.3.

Table 9.1 presents a breakdown of estimated costs for a field test using the standardized interagency work breakdown structure (WBS) as described in the Federal Remediation Technologies Roundtable (FRTR) document. Before- and after-treatment costs are shown at the second level of the WBS and costs directly associated with treatment are shown at the fifth level of the WBS.

9.2 ESTIMATED FULL-SCALE COSTS

The total cost of implementing the full-scale design presented in Chapter 7 is estimated at approximately $900,000 for 45,000 yd^3 ($688,000 for 34,400 m^3), or $20 per cubic yard ($15 per cubic meter) and $64,000 per acre ($26,000 per hectare). This estimate is based on the conditions at the Craney Island facility and operation of the system for two growing seasons. If additional time is required to reach an acceptable end point, the estimated total cost would be $110,000 per year or $2.4 per cubic yard ($1 per cubic meter) or $7900 per acre ($3200 per hectare) per year.

Table 9.2 presents a breakdown of estimated cost for a field test using the standardized WBS as described in the FRTR document.

9.3 ECONOMIC ANALYSIS

This section provides an economic analysis of phytoremediation of hydrocarbon contaminants in soil. A full-scale cost estimate model is developed and a detailed design and cost

Table 9.1 Craney Island Phytoremediation Field Test: Estimated Costs

Interagency WBS #	Cost Element	Unit Cost ($)	No. of Units	Cost[a] ($)
Before-treatment costs				
	Project management • Work plan preparation • Initial meetings (Note: This element is not listed in the WBS)	76,000	Lump sum	76,000
33 01	Mobilization and preparatory work • Mobilization of equipment, material, and personnel; included initial site visits and mobilization of equipment such as the irrigation piping	8,800	Lump sum	8,800
33 02	Monitoring, sampling testing, and analysis • Treatability studies: plant selection field test	10,000	Lump sum	10,000
33 03	Site work • Site preparation including clearing and grubbing was completed by the Craney Island facility prior to the start of this project	0	Lump sum	0
Treatment cost elements				
33 11	Solids preparation and handling • Excavation, screening, and spreading of contaminated soil were completed by the Craney Island facility prior to the start of this project	0	Lump sum	0
33 11	Mobilization/setup • Installation of the treatment cell, including grading, geogrid/synthetic liner/sand layer, and berms, was completed by the Craney Island facility prior to the start of this project	0	Lump sum	0
33 11	Equipment and fixed costs • Purchase and installation of irrigation system	3,000	Lump sum	3,000
33 11	• Purchase and installation of soil pore water collectors (lysimeters)	2,000	Lump sum	2,000
33 11	• Generator for electrical power	2,000	Lump sum	2,000
33 11	• Monitoring equipment costs	5,000	Lump sum	5,000
33 11	Capital costs • No capital costs were expended for the phytoremediation field test project (capital costs defined as purchase of assets with a useful life of more than 1 year); the existing Craney Island biotreatment facility was used	0	Lump sum	0
33 11	Consumables costs • Fertilizer	500	Lump sum	500
33 11	• Laboratory chemicals	500	Lump sum	500
33 11	• On-site utilities: irrigation water	1,000	Lump sum	1,000
33 11	Supplies costs • Miscellaneous supplies	500	Lump sum	500
33 11	Treatment labor requirements and costs • Estimated labor required for seeding, tending, and harvesting phytoremediation plots	384/day	96 days	37,000

Table 9.1 Craney Island Phytoremediation Field Test: Estimated Costs (continued)

Interagency WBS #	Cost Element	Unit Cost ($)	No. of Units	Cost[a] ($)
33 11	• Estimated additional labor associated with research aspects of the phytoremediation field test	784/day	96 days	75,000
33 11	Health and safety plan implementation costs • Health and safety consumables	50/day	96 days	4,800
33 11	Sampling, analytical and other measurement costs • Sampling labor	384/day	24 days	9,200
33 11	• Analytical costs (in-house gas chroma-tography and outside lab costs) price includes bottles, shipping, lab reports	100 200 400	500 samples 50 samples 50 samples	50,000 10,000 20,000
After-treatment cost elements				
33 21	Site demobilization costs	4,400	Lump sum	4,400
Total				320,000

[a] Costs rounded to two significant figures.

estimate is provided for full-scale phytoremediation at the Craney Island facility. Scale-up, process design, and financial issues are addressed. Finally, estimated costs are compared to costs for alternative technologies.

9.3.1 Cost Model

A full-scale cost model was developed, using the process design of the Craney Island field test and prices obtained from vendors and cost elements found in standard cost-estimating guides. The purpose of the cost model is to illustrate the relationships among the cost elements and to provide, at best, order-of-magnitude cost estimates. The cost values are estimates of mean costs over a wide range of possible costs and should not be used to predict costs at a particular site.

Figure 9.1 presents the estimated total costs for phytoremediation systems operated with irrigation for 2 years for areas ranging from 3 to 27 acres (1 to 11 ha). Cost spreadsheets, cost curves, and references are provided in Appendix 3. By expressing costs as a function of the operating time frame, the costs become independent of soil and contaminant characteristics (for which there is not yet a design relationship established). A soil with a higher contaminant concentration would be assumed to require a longer treatment period. An example of the relationship between total cost and operating period is presented in Figure 9.2.

9.3.2 Process Design Assumptions and Other Bases for Costs

The following three major assumptions are inherent in the cost estimate model and in evaluation of full-scale systems:

• The requirements and limitations described in Section 10.2 are addressed by assuming that no limitations to scale-up exist. No costs for the additional research and development critical to the implementation of this technology have been added to the cost estimates. An initial plant selection study is included. The water limi-tation is addressed by assuming adequate water sources exist on the site. The costs

Table 9.2 Full-Scale Costs: 14-Acre (5.67-Ha) System, 2 Years

Interagency WBS #	Cost Element	Unit Cost ($)	No. of Units	Cost[a] ($)
Before-treatment costs				
	Project contracts and engineering • Work plan preparation • Initial meetings (Note: This element is not listed in the federal WBS)	76,000	Lump sum	76,000
33 01	Mobilization and preparatory work • Mobilization of equipment and material	4,400	Lump sum	4,400

33 02	Monitoring, sampling, testing, and analysis
	Initial site investigation/characterization
	For each acre:
	  Labor $500
	  Travel $200
	  5 TPH × $200
	  5 PAH × $400
	  1 agricultural parameters (pH, CEC, etc.)
	   × $300
	  Supplies/personal protection equipment $100
		4,100	14 acres	58,000

	Plant selection study	10,000	Lump sum	10,000
33 03	Site work and soil spreading • Site preparation including screening, spreading, and additional tilling of soil	470,000	Lump sum	470,000
Treatment cost elements: 2 years of operation				
33 11	Equipment and fixed costs • Purchase and installation of irrigation system	76,000	Lump sum	76,000
33 11	• Hydroseeding	1,850	14 acres	26,000
33 11	• Harvesting, fertilizing, field costs	37,000	2 years	74,000
33 11	Consumable costs • On-site utilities: irrigation water	0.10	23,000 TGAL	2,300
33 11	Professional labor requirements and costs • Estimated labor required for tending and sampling phytoremediation plots	460	48 days	22,000
33 11	Health and safety plan implementation costs • Health and safety consumables	150	48 days	13,000

33 11	Analytical and other measurements costs
	For each acre:
	  2 per year × 3 TPH × $200
	  2 per year × 3 PAH × $400
	  2 agricultural parameters (pH, CEC, etc.)
	   × $300
		4,200	14 acres	59,000

After-treatment cost elements				
33 21	Site demobilization costs	4,400	Lump sum	4,400
Total				895,100

[a] Costs rounded to two significant figures.

of commercially available water or water source development will vary widely from location to location and from region to region within the U.S.

• Hydrocarbon contaminant concentrations are assumed to be moderate and not toxic to plants. Other constituents in the soil are assumed to not be toxic to plants and to not define regulatory end points.

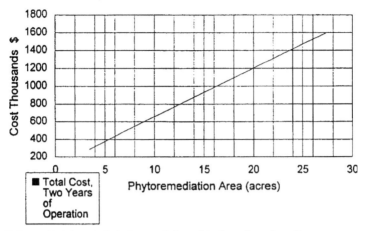

Figure 9.1   Estimated total cost of phytoremediation of hydrocarbons in soil.

- The depth of soil that can be effectively treated is assumed to be 2 ft (0.6 m). Greater depths are anticipated to be possible for treatment, but existing data are insufficient for design and cost estimates.

9.3.3 Construction and Equipment Costs

The major construction costs consist of creation of a bermed treatment area and installation of irrigation equipment for locations requiring irrigation. Water resource development will be a major cost at some arid sites. However, the costs for water resource development were too regionally variable to estimate for this report. Synthetic liner installation for sites requiring prepared-bed systems will also be a major cost.

The total cost for a 14-acre (5.7-ha) system operating and requiring an irrigation system and a bermed area but not requiring a water production well or synthetic liner is estimated to be $94,000. These costs are presented in Appendix 3.

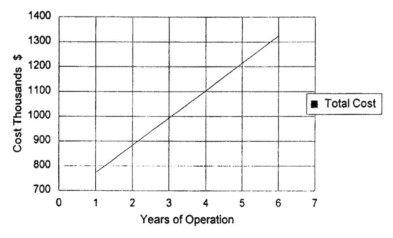

Figure 9.2   Total costs of phytoremediation of hydrocarbons in soil for 14-acre system over several years of operation.

9.3.4 Other Fixed Capital Costs, Including Plant Species Selection

Other fixed capital costs are basic start-up costs common to most systems. These include the costs of design including greenhouse plant selection studies, permitting, mobilization, site preparation, soil spreading and preparation, and seeding. Site preparation costs assume that soil is present at the surface or in stockpiles and that site investigation and excavation of contaminated soils were conducted previously.

9.3.5 Operating Costs

Operating costs comprise all of the costs which must be paid repeatedly throughout the life cycle of the phytoremediation program. These include operational and environmental monitoring, fertilizing, harvesting, and project management and reporting. Operating costs for a 14-acre (5.7-ha) system were estimated at $110,000 per year.

9.3.6 Financial Considerations

9.3.6.1 Rate of Fixed Capital Depreciation

Typical phytoremediation of soil may require only a very low level of capital asset expenditure (i.e., assets that have a service life of more than 1 year and for which depreciation is allowed by the Internal Revenue Service). For a typical, 14-acre (5.7-ha) phytoremediation system which uses irrigation but does not require a water supply well or a synthetic liner, less than 10% of the total cost is for capital asset costs, the irrigation system. The cost estimate is therefore not very sensitive to the rate of capital depreciation. The cost estimate model presented in this section involved capital expenditures only for the irrigation system and the optional lined bed installation. At the scale of most phytoremediation applications (less than 100 acres) (less than 40.5 ha), the other agricultural equipment required is most economically leased or contracted as a service, similar to custom farming operations.

9.3.6.2 Time Value of Money

Phytoremediation will usually involve a start-up period with capital expenditures, followed by several years of operational costs. To convert the operational costs into an estimated equivalent present value, the time value of money (commonly, the interest rate) must be used to discount future costs. Because the period of time will be less than 10 years for most systems, the total cost will be relatively insensitive to variation in the discount rate. For example, using an interest rate of 5%, the net present value of a 14-acre system operating for 2 years is estimated by discounting the second year of operations costs: $110,000 \times 0.9524 = \$103,000$. An interest rate of 10% would result in a discounted second-year operations cost of $100,000. The difference of $3000 represents 0.3% of the total cost of $900,000.

9.3.6.3 Sensitivity of Estimated Cost to Gross Margin and Profit Margin

Phytoremediation costs will have average sensitivity to gross margin and profit margin. The phytoremediation technology evaluated in this book does not rely on exotic materials, difficult to obtain equipment, or patented processes.

The total final cost of phytoremediation at a particular site will be highly dependent upon the following factors:

- *Scale* — The cost per cubic yard of treated soil will decrease with increasing scale. The cost for a 3.4-acre (1.4-ha) system was estimated at $26 per cubic yard ($20 per cubic meter) whereas the cost for a 14-acre (5.7-ha) system was estimated at $20 per cubic yard ($15 per cubic meter). These estimates include capital plus 2 years of operations.
- *Soil handling* — The estimates in this book included spreading of stockpiled soil. Soil handling such as excavation, stockpiling, and spreading will contribute $5 to $10 per cubic yard ($4 to $8 per cubic meter). Thus, if phytoremediation can be accomplished *in situ* without excavation and spreading, important cost savings can be realized.
- *Initial and final concentration and end points* — With all other factors equivalent, a site with higher initial concentrations or lower final concentration and end point requirements will require a longer treatment period and, hence, higher costs. The annual cost of operations will be at least $2 to $4 per cubic yard ($1.5 to $3 per cubic meter).

9.3.7 Comparison of Project Costs to Cost for Demonstrations of Similar Technology

No comparable costs were available for demonstrations of phytoremediation technologies. Other field demonstrations of phytoremediation either did not have published cost data or were not comparable because they used very different plant systems, such as hybrid poplar and willow (*Salix*) species.

9.3.8 Comparison of Project Costs to Cost for Alternative Treatments

As discussed in Section 9.1, the project costs associated with the Craney Island phytoremediation field test do not accurately represent the unit costs of this technology. Although it provides only an order-of-magnitude estimate at best, the cost estimate model of full-scale systems is a more accurate representation and is used in this section for unit cost comparisons with other technologies.

Cost comparisons among alternative treatment technologies are difficult because of differences in scale and the variability in the contaminated soil matrix and the treatment end points. In addition, it is often not clear what elements are included in the reported unit costs.

The closest match to the performance and application of this phytoremediation process is solid-phase biological treatment of hydrocarbon-contaminated soil. In this process, soil is spread onto a treatment area and periodically tilled, watered, fertilized, and limed as required to promote microbial degradation of the hydrocarbons. Solid-phase biotreatment costs vary widely depending on the scale, matrix, and end points. However, for a moderate-sized site of 11,000 yd^3 (8,400 m^3), a unit cost of $30 per cubic yard ($23 per cubic meter) is typical (see Appendix 3). This compares with a unit cost of about $20 per cubic yard ($15 per cubic meter) for phytoremediation, assuming a 2-year time frame to reach acceptable end points for phytoremediation and a one-season time frame for solid-phase biotreatment. The site characteristics that may be most favorable to implementation of phytoremediation are discussed in Chapter 10.

CHAPTER **10**

Performance and Potential Application

10.1 PERFORMANCE

This section discusses the performance of the Craney Island phytoremediation system and relates this performance to specific factors and design parameters that will affect phytoremediation of hydrocarbons in soil. Phytoremediation was shown in the Craney Island field test to result in only moderate decreases in hydrocarbon contaminant concentrations in its 2-year duration. However, the performance of a remediation system consists of not only the rate and extent of treatment as measured by numerical standards, but also includes other effects of the process which decrease the potential for short-term or long-term impacts to potential receptors. A unique attribute of phytoremediation is its potential for promoting the ecological restoration of land. Although ecological restoration was beyond the scope and time frame of the Craney Island field test, this section addresses ecological restoration as another measure of performance which may be applicable to future full-scale sites.

10.1.1 Maximizing Treatment Rate

This section discusses treatment rates and the factors which were important to maximizing the treatment rate in the Craney Island phytoremediation field test. Although the term "treatment" in the broadest sense includes reduction in mobility and toxicity as well as reduction in total chemical concentrations, "treatment" will be used in this section to refer to reduction in chemical concentrations.

Several important factors which affect treatment rate are inherent in the contaminated soil matrix and are thus beyond the control of the phytoremediation process. The rate and extent to which treatment will occur in hydrocarbon mixtures are highly dependent on the characterization of the mixture. The extent of total petroleum hydrocarbon (TPH) degradation (as expressed as a percentage of the final vs. initial concentration) will be less in weathered mixtures in which the readily degraded, lower molecular weight hydrocarbons have already been removed. The rate and extent of treatment also appear to depend on the natural physical and chemical characteristics of the soil. The organic carbon content and the soil texture play significant roles (Linz and Nakles, 1997).

Phytoremediation of hydrocarbons in soil is characterized by slower rates of treatment at lower cost in comparison to remedial alternatives such as thermal desorption and slurry biotreatment. The best treatment rate observed in the Craney Island field test, about 50% TPH dissipation over 22 months, was lower than the rate expected for more active solid-phase biological treatment by soil tilling. A typical solid-phase biological treatment system would result in at least 80% dissipation of moderately weathered diesel-range TPH over the same time period.

The choice of plant species and establishment of active microbial communities appeared to be significant factors affecting treatment rates in the Craney Island field test. The extent of TPH degradation was not simply directly proportional to plant growth or root density. The clover species exhibited the greatest TPH degradation and yet had the lowest root and shoot biomass. Its roots were coarser and shallower than the other species. Optimum maintenance of the clover would have required more irrigation than the other species. One mechanism proposed for this in the final report of the field test was that dying clover roots provided a substrate that encouraged greater microbial activity and cometabolism of hydrocarbon contaminants. Thus, the species providing the highest degradation rate may not be the species with the highest growth rates or the species which performs best under low-maintenance conditions.

Biodegradation by active microbial populations appears to be the primary mechanism for decrease in TPHs and polycyclic aromatic hydrocarbons (PAHs) in this field test. The presence of vegetation most likely affects degradation indirectly by providing an enhanced chemical, biological, and physical environment which promotes microbial action. Thus, careful and frequent monitoring of soil environmental conditions including moisture content, soil air, pH, nitrogen and phosphorous levels, and temperature will be critical to the performance of phytoremediation of hydrocarbons in soil.

Other factors of importance were directly related to maintaining plant and microorganism growth. Maintaining optimum moisture conditions by draining excess water and by irrigating during dry periods was an important activity. Maintaining nutrient levels by addition of fertilizer was also important, as was harvesting at the appropriate periods. Root growth has been related to addition of fertilizer and lime (Tilsdale and Nelson, 1975). The appropriate fertilization rate for nonlegumes without additional microbial activity from metabolism of hydrocarbons is C:N:P = 100:20:10. Nitrogen fertilization at this rate may inhibit root nodules forming on legumes such as clover (Banks et al., 1998).

10.1.2 Minimizing Impacts to Potential Receptors

The beneficial effects of phytoremediation in soil may include not only decreased chemical concentrations but may also include decreased availability and toxicity and decreased potential for migration of contaminants and contaminated soil particles from the site. The goal of these forms of treatment is to minimize the impacts to potential receptors in order to satisfy the short-term and long-term objectives of a site remediation plan.

Although the degree of hydrocarbon availability and the toxicity of the soils were not measured in the Craney Island field test, it is known that the organic carbon content of the soil increased substantially in the phytoremediation plots, primarily through the growth and sloughing of plant roots and the incorporation of dead plant shoots into the soil surface. The organic carbon content of the soil has been related to the availability (extent of sorption) of PAHs and other hydrophobic organic chemicals (Linz and Nakles, 1997). The increased organic carbon in the phytoremediation plots may have led to increased sorption of PAHs and, therefore, reduced bioavailability and potential toxicity. Thus, the phytoremediation process may have had beneficial treatment effects beyond reduction of chemical concentrations.

The phytoremediation process may also have resulted in reducing migration of soil from the site. The plant growth promoted in phytoremediation systems acts to physically stabilize the soil and to lessen the transport of soil caused by wind and rainfall. The use of plant cover to control erosion is a well-established practice. The Universal Soil Loss Equation, used by the U.S. Department of Agriculture to estimate soil loss, recognizes the importance of vegetation cover. Covers such as grasses reduce rainfall energy, and plant roots and mulches protect the soil from erosive forces. Grass crops and grass pasture are characterized as among the best covers for minimizing soil erosion, second only to woodlands (Novotny and Chesters, 1981).

10.1.3 Maximizing Ecological Restoration

Ecological restoration may be an objective of a site remediation plan which could be addressed by phytoremediation. The Craney Island phytoremediation system was a short-term field test primarily concerned with the use of plants to enhance the dissipation of chemical concentrations in the soil. No ecological surveys were conducted in the planted and unplanted field test plots, and no firm conclusions can be drawn with regard to ecological impacts within the treatment area. However, the positive response of plants grown under field conditions in hydrocarbon-contaminated soil indicated the potential for ecological benefits.

10.1.4 Site Characteristics and Operating Parameters Affecting Cost and Performance

Operating parameters affecting cost and performance in the phytoremediation system are listed in Table 10.1, in accordance with the Federal Remediation Technologies Roundtable (1995) format. Measurement procedures and potential effects on cost or performance are also listed.

10.2 SCALE-UP REQUIREMENTS AND LIMITATIONS

The critical scale-up requirement and limitation for this technology is the need for more information regarding the mechanisms, performance, and economics of phytoremediation of hydrocarbon-contaminated soils. Insufficient data exist at present to provide predictive relationships upon which a rational full-scale design can be based. For example, the design of a full-scale system with conditions identical to the Craney Island field test, but with a different soil type with more clay, would not be able to accurately predict the rate or extent (end point) achievable by the system.

Other scale-up requirements and limitations that are of economic importance include water sources for irrigation, sources of large quantities of the specified plant seeds or cuttings, and the greater variability of soil and contaminant characteristics to be encountered at larger sites. The following paragraphs discuss these limitations.

10.2.1 Water

Water sources for irrigation may be required for many locations. Even temperate climates, such as at the Craney Island facility in Virginia, experience dry summer periods which may adversely affect performance. While small-scale systems will usually be able to rely on existing local water supplies, the cost of developing water sources for scaled-up phytoremediation systems may limit their size. However, not enough data currently exist to provide a quantitative relationship between rainfall and remediation performance to enable an economic evaluation of the need for irrigation.

10.2.2 Seeds

A second requirement of scaled-up systems will be reliable sources of the particular plant seeds or cuttings required for phytoremediation. This requirement may limit the scale of operations unless common plant varieties are effective.

Table 10.1 Operating Parameters: Measurement Procedures and Potential Effects on Treatment Cost or Performance

Operating Parameters	Measurement Procedures[a]	Important to Document Measurement Procedure?	Potential Effects on Cost or Performance
Moisture content	Soil moisture content is typically measured using a gravimetric ASTM standard, D2216-90, Test Method for Laboratory Determination of Water (Moisture) Content of Soil and Rock, also ASA No. 9 21-2.2.	No	The moisture available to plants is a critical matrix characteristic for plant selection.
pH	EPA SW-846 Method 9045 and ASTM methods for soil (ASTM D 4972-89, Test Method for pH of Soils) and groundwater (ASTM D 1293-84).	No	The pH of the matrix can impact the solubility of contaminants and all biological activity; it can also affect the operation of phytoremediation.
Residence time	Residence time is the amount of time that a unit of material is processed in a treatment system. Residence time is measured by monitoring the length of time that a unit of soil is contained in the treatment system.	No	Residence time is important to measure during treatment.
Temperature	Temperature is measured using a thermometer or thermocouple.	No	Temperature affects the rate of biological activity.
Precipitation	Rain gauge recorded over time periods of 1 week is often sufficient.	No	Water is essential to plant growth and microbial activity. Excess water may require water management.
Irrigation rate	Rain gauge.	No	Water is essential to plant growth and microbial activity.
Microbial concentration	The number of microorganisms per unit volume in a matrix. Part 9000 of Standard Methods for the Examination of Water and Wastewater contains many methods for laboratory analysis. Portable test kits are available for field use.	Yes	Microbial biomass concentration is an important parameter for biodegradation and bioremediation. Microorganisms are necessary to effect treatment.
Microbial activity, oxygen uptake rate, CO_2 production, BIOLOG	Microbial activity is measured to assess hydrocarbon degradation. Oxygen uptake rate is measured using ASTM D 4478-85, Standard Test Methods for Oxygen Uptake. CO_2 evolution can be measured with a CO_2 monitor. The BIOLOG system is sometimes used to characterize activity.	Yes	Microbial activity is an important parameter for phytoremediation of hydrocarbons in soil.
Root and shoot growth	Measurement of plant root and shoot mass per volume of soil using standard agronometric methods.	Yes	Plant biomass is an important parameter in phytoremediation, indicating the vigor and depth of growth.

Parameter	Method	Applicable	Significance
Ecological restoration	Standard ecological survey techniques.	Yes	The health of the overall ecosystem may be important to assessing the performance of some phytoremediation systems.
Nutrients and other soil amendments	N, P, and trace inorganics. Typically reported as C:N:P. Carbon, as total organic carbon, measured with EPA SW-846 Method 906 or Agronomy Method ASA No. 9 33-6. Nitrogen measured as ammonia N with ASTM D 1426-89, and as nitrite–nitrate with ASTM D 3867-90. Phosphorusis measured using ASTM D 515-88. Calcium and magnesium are measured with ASTM D 511-88.	Yes	Nutrients affect the rate of biological activity and therefore contaminant biodegradation.
Contaminants in soil, leachate, and plant biomass	TPH, PAH, and other constituents of interest may be measured by standard methods (see Table 8.2).	Yes	A time series of these measurements indicates the progress toward contaminant dissipation and allows a mass balance to document contaminant fate.
Leachable hydrocarbons	The Synthetic Precipitation Leach Procedure EPA Method 1312, column tests, or slurry availability tests.	Yes	A time series of these measurements indicates the progress toward decreased contaminant availability to potential receptors.
Toxicity	Earthworm toxicity and surrogate toxicity measurements using polymetric fibers.	Yes	A time series of these measurements indicates the progress toward reduced contaminant toxicity.

[a] ASA = American Society of Agronomy, ASTM = American Society for Testing and Materials, EPA = U.S. Environmental Protection Agency.

10.2.3 Variability

Larger systems will tend to include more varied soil types and contaminant concentrations. This variability will require larger systems to have flexible treatment regimes and time frames. The economics of operating a large system over a long period of time in order to finish treatment of a fraction of the more contaminated soils or soils that are characterized by more difficult conditions, such as clayey soils, may limit the scale of phytoremediation systems or require flexible and carefully designed subsystems.

10.2.4 Techniques and Equipment

These are not anticipated to be scale limiting or to present unusual requirements because phytoremediation uses conventional agricultural techniques and equipment which are available in most areas of the U.S.

10.3 APPLICABLE CONTAMINANTS AND CONCENTRATION RANGES

This section examines the potential application of phytoremediation of hydrocarbon-contaminated soils. Applicable contaminants and concentration ranges are identified, suitable site characteristics are discussed, and the technology is evaluated within the current U.S. regulatory context.

There are insufficient data regarding the performance of phytoremediation to define the full range of potential contaminants and concentrations which could be remediated by this technology. However, the results of recent studies, including the Craney Island field test, a recent work sponsored by the Petroleum Environmental Research Forum/Gas Research Institute (Drake, 1997), and others, suggest that phytoremediation may be applicable to a broad range of hydrocarbon mixtures.

The Craney Island field test and other work indicate that hydrocarbons at concentrations commonly found at contaminated sites have little or no toxic effects on the common grasses, legumes, and trees used in these tests. Most of these tests included an initial germination study to select plants that thrived in these soils. More important toxic effects may be related to the salinity of contaminated soils, which may limit the use of some plant species and may limit the applicability of phytoremediation of high concentrations of contaminant mixtures such as oily brines from exploration and production sites.

The categories of hydrocarbon mixtures commonly found at sites that have the most economic importance include:

- *Crude oils* — Crude-oil-contaminated sites include oil pipeline and storage spill sites and oil exploration and production sites. The characteristics of crude oil mixtures vary widely with the source.
- *Gas pit residuals* — Oily gas pit residuals are found at natural gas exploration and production sites.
- *Refined petroleum products, including fuel oils, diesel engine fuels, gasoline, and jet fuels* — These mixtures are encountered in soil at refining, storage terminal, and marketing facilities of the oil industry and at numerous federal sites.
- *Coal tar residuals found at former manufactured gas plant sites and wood-treating sites*

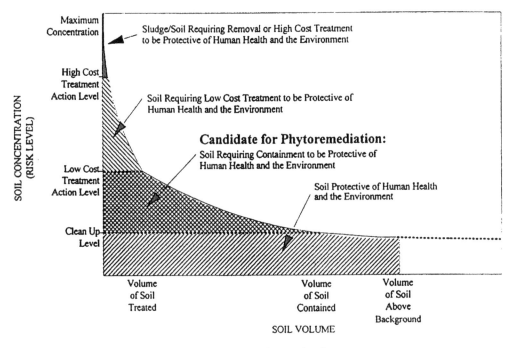

Figure 10.1   Relationship between soil volume and cleanup levels.

Within each of these categories, the hydrocarbon contaminant mixtures are often characterized by the predominant molecular weight range of aliphatic and aromatic hydrocarbon compounds. Although there is a general inverse relationship between molecular weight and biodegradability in soil, there are not enough data to quantify performance relationships for phytoremediation.

Phytoremediation is a slow-rate, low-cost technology that is anticipated to be most applicable to the large volume of soil containing low concentrations of contaminants. Figure 10.1 illustrates the general relationship between contaminant concentration and soil volume at contaminated sites. A relatively small volume of soil and sludge contains very high contaminant concentrations and may require removal or high-cost destructive treatment. A moderate volume of soil contains an intermediate range of concentrations which require lower cost treatment or off-site disposal. The largest volume of soil contains contaminant concentrations which pose no immediate threat to potential receptors but are above acceptable long-term environmental end points and require containment or very low-cost remedial action. This large volume of soil represents the most applicable concentration range for phytoremediation.

10.4 RANGES OF SITE CHARACTERISTICS SUITABLE FOR PHYTOREMEDIATION OF HYDROCARBONS IN SOIL

Phytoremediation of hydrocarbons in soil is potentially limited only by the ability to establish and maintain viable plant cover. The importance of species selection and consideration of native or locally adaptive species cannot be overemphasized. The important site

characteristics to be taken into account when considering phytoremediation include temperature, latitude, precipitation, slope, soil characteristics, and depth of contamination.

10.4.1 Temperature/Latitude

The regional location of the site will obviously affect the length of the growing season
and have critical impact on the feasibility of implementing phytoremediation. Overwintering
strategies and use of perennials are important considerations for northern climates. The
quantitative relationship among ambient temperature, latitude, and remediation performance
for various species has not yet been established.

10.4.2 Precipitation

The implementation of phytoremediation in arid regions would require special plant
selection and is anticipated to be less economically feasible than in climates where little or
no irrigation is required. Especially wet site conditions, such as inundation of the site, would
require special plant selection and could preclude effective phytoremediation. The quantitative relationship between precipitation and remediation performance for various species has
not yet been established.

10.4.3 Slope

Although sites with extreme slopes may preclude effective phytoremediation, the technology could be applied to sites with contaminated soil located on hillsides. Phytoremediation
using broadcast seeding and little or no harvesting might provide a distinct advantage over
alternative *in situ* technologies such as tilling, which would not be feasible on steep slopes
or could cause excessive soil loss due to erosion. Phytoremediation systems could be designed to minimize erosion by following erosion-control practices such as contouring and
terracing.

10.4.4 Soil Characteristics

Salinity, texture, and other soil characteristics are important considerations, as discussed
in Chapter 8. Highly saline soils or highly clayey soils may require special plant selection or
soil amendments. The quantitative relationship between soil characteristics and remediation
performance for various species has not yet been established.

10.4.5 Depth of Contamination

Phytoremediation is most applicable to sites with contaminated soils at depths of at least
2 ft (0.6 m). The maximum effective depth and the relationship between depth and remediation
performance for various species have not yet been established (Aprill and Sims, 1990).

10.5 REGULATORY REQUIREMENTS FOR
PERFORMANCE AND COMPLIANCE CRITERIA

Performance requirements that define the end points to be reached and compliance criteria
that ensure safe and environmentally benign operations will have profound effects on the
application of phytoremediation technology. This section discusses the implementation of

phytoremediation of hydrocarbon-contaminated soils in the context of the current and emerging regulatory context.

10.5.1 Regulatory Performance Requirements

Regulatory performance requirements for sites with contaminated soils currently emphasize numerical cleanup standards. An emerging body of scientific evidence and site experience suggests that a regulatory framework for many sites in the future will be a risk-based approach to site management based on the actual availability of chemicals in the soil. Moreover, remedial performance objectives will more frequently include measures of ecological restoration. Phytoremediation has great potential to play an important role in this emerging regulatory context.

Remediation end points for petroleum-hydrocarbon-contaminated soil are currently most often expressed as numerical cleanup standards. These vary from state to state within the range of 100 to 10,000 mg/kg TPH. Specific cleanup standards for PAHs at these sites may also apply. Coal-derived hydrocarbon contaminants in soil at manufactured gas plant sites are usually regulated using a similar numerical approach. Phytoremediation has been demonstrated to result in decreased hydrocarbon contaminant concentrations and may be effective in achieving numerical cleanup standards at sites with contaminant concentrations close to the cleanup standards.

An emerging, additional framework for achieving site closure uses a site-specific risk-based approach to define environmentally acceptable remediation end points. For sites with hydrocarbon-contaminated soil, this approach is based in large part on research examining the bioremediation of hydrocarbons in soil. The following results of that research are important to understanding the application of phytoremediation in this regulatory context:

- Availability limits biodegradation. Biodegradation proceeds to a concentration which then does not decrease, or which decreases only very slowly, as treatment operations are continued. Further decreases in concentration are limited by the availability of the soil-bound hydrocarbons to the microorganisms.
- Hydrocarbons that remain after initial biodegradation have much lower aqueous leachability and are less available to other organisms than was the hydrocarbon mixture present in the soil prior to biodegradation.
- Hydrocarbons that have simply been in contact with soil for a period of time (aged), even without biodegradation, are less available to many organisms and have reduced aqueous leachability than hydrocarbons that have been in contact with soils for a shorter period.

These results suggest that bioremediation, including phytoremediation, can decrease hydrocarbon concentrations such that the hydrocarbons remaining in the soil no longer pose an unacceptable risk to the environment or human health. These residual hydrocarbons would no longer leach to groundwater, harm ecological receptors, or be available to humans. This concept of environmentally acceptable end points and the supporting research were presented in a recent publication (Linz and Nakles, 1997).

As research on phytoremediation of hydrocarbons in soil continues, it may be shown that plants effectively modify the soil matrix by increasing soil organic matter and through other mechanisms in ways that further lead to environmentally acceptable end points.

This approach is currently being examined in the context of risk-based cleanup action (RBCA) initiatives. The research supporting the concept of environmentally acceptable end points could be used a to define a multitiered RBCA assessment such as that developed by the American Society for Testing and Materials.

Enhancement of ecological restoration is becoming more important as a management objective for some sites. Phytoremediation may serve an important role in addressing ecological risk and restoration. It could be integrated with ecological management practices to promote the establishment of desirable plant and animal species, leading to natural succession and biodiversity at the site.

10.5.2 Operational Compliance Criteria

Operational compliance criteria are short-term criteria imposed on the remediation system during its operation. These include measures to protect public health and the environment during phytoremediation and measures for the protection of workers conducting the phytoremediation operations.

The operational compliance criterion that will have the most profound effect on the application of phytoremediation of hydrocarbons in soil is the protection of groundwater during the process. It will be necessary to demonstrate that either the hydrocarbons in the soil do not pose a significant threat to groundwater or that the risk is acceptable over the time period of the phytoremediation process and phytoremediation will result in soil that no longer poses a significant threat to groundwater. If protection of groundwater requires the construction of a prepared-bed system with an impermeable synthetic liner, then the cost of phytoremediation will increase dramatically and may render the technology infeasible at particular sites.

Other operational compliance criteria include air quality monitoring and control to ensure that there are no impacts due to fugitive emissions of vapors or particles from the system. Runoff monitoring and control may also be important criteria to consider during installation.

Aesthetics considerations are readily addressed by phytoremediation, which has as an intangible benefit the improvement of site aesthetics as provided by the green plant cover.

Worker health and safety considerations include standard OSHA requirements for protection of workers from hazards posed by dermal contact with contaminated soil and by noise and air quality monitoring and control.

CHAPTER **11**

Design/Evaluation References and Bibliography

11.1 REFERENCES

Aprill, W. and Sims, R.C. 1990. Evaluation of the use of prairie grasses for stimulating PAH treatment in soil, *Chemosphere*, 20:253–266.

Banks, M.K., Schwab, A.P., and Govindaraju, R.S. 1996. Phytoremediation of Soil Contaminated with Hazardous Chemicals: Monthly Status Report, September 1996, prepared for AATDF.

Banks, M.K., Schwab, A.P., Govindaraju, R.S., Kulakow, P., Rathbone, K., and Butler, S. 1998. Phytoremediation of Soil Contaminated with Hazardous Chemicals: Final Technical Report, prepared for AATDF.

Drake, E.N. 1997. Phytoremediation of aged petroleum hydrocarbons in soil, presented at IBC's 2nd Annual Conference on Phytoremediation, June 18 and 19, IBC, Seattle, WA.

Federal Remediation Technologies Roundtable. 1995. Guide to Documenting Cost and Performance for Remediation Projects, Member Agencies of the Federal Remediation Technologies Roundtable.

Linz, D.G. and Nakles, D.V., Eds. 1997. *Environmentally Acceptable Endpoints in Soil*, American Academy of Environmental Engineers, Annapolis, MD, 217–221.

Novotny, V. and Chesters, G. 1981. *Handbook of Nonpoint Pollution*, Van Nostrand Reinhold, New York.

Schwab, A.P., Su, J., Wetzel, S., Pekerak, S., and Banks, M.K. 1998 (submitted). Extraction of petroleum hydrocarbons from soil by mechanical shaking, *Environ. Sci. Technol.*

Tisdale, S.L. and Nelson, W.L. 1975. *Soil Fertility and Fertilizers*, 3rd ed., Macmillan, New York, 502.

U.S. EPA. 1981. Process Design Manual Land Treatment of Municipal Wastewater, USEPA 625/1-81-013, U.S. Environmental Protection Agency, Cincinnati, OH, E-15–E-22.

11.2 BIBLIOGRAPHY

The following general texts may be helpful in understanding the fundamentals of phytoremediation of hydrocarbons.

Atlas, R.M. 1984. *Petroleum Microbiology*, Macmillan, New York.

Bohn, H.L., McNeal, B.L., and O'Connor, G.A. 1979. *Soil Chemistry*, John Wiley & Sons, Canada.

Brady, N.C. 1974. *The Nature and Properties of Soils*, 8th ed., Macmillan, New York.

Dragun, J. 1988. *The Soil Chemistry of Hazardous Materials*, The Hazardous Materials Control Research Institute, Rockville, MD.

Kostecki, P.T. and Calabrese, E.J. 1991. in *Hydrocarbon Contaminated Soils and Groundwater*, Bell, C., Ed., Lewis Publishers, Boca Raton, FL.

Loehr, R.C., Jewell, W.J., Novak, J.D., Clarkson, W.W., and Friedman, G.S. 1979. *Land Application of Wastes*, Vol. 1, Van Nostrand Reinhold, New York.

Appendices

Kansas State University Methods for Standard Analysis of Total Recoverable Petroleum Hydrocarbons in Soils Using Infrared Spectrophotometry

1 OBJECTIVES

1.1 This method is modified from EPA Method 418.1 (modified) and SW846 Methods 3540A and 3630A. It is to be used for the measurement of fluorocarbon-113- (Freon) extractable petroleum hydrocarbons from soils/solids.

1.2 The method is applicable for measurement of light fuels, although loss of about half of any gasoline present during the extraction can be expected.

2 APPROACH

2.1 Soil samples are extracted using sonication. A 30-g sample is mixed with anhydrous sodium sulfate to form flowable material, and then 100 ml Freon is added. The mixture is sonicated and filtered. Interferences are removed with silica gel absorbent. The analysis of the extract is performed by direct comparison with standards.

3 POTENTIAL INTERFERENCES

3.1 Solvents, reagents, glassware, and other processing hardware may result in interferences to sample analysis. All of these materials must be demonstrated to be free from interferences under the conditions of the analysis by analyzing method blanks. At least one blank must be present in every set of samples (regardless of the number of samples), and the minimum rate of blank analysis is one per 20 samples. Specific selection of reagents may be required.

3.2 Soap residue on glassware may cause degradation of certain analytes. Follow the glassware cleaning protocol in Section 6.4.

4 MATERIALS

4.1 *Soxhlet extraction apparatus* — A standard soxhlet extractor is used with 24.40 or 45/50 ground glass joints, 500-ml boiling flask, heating mantle, and condenser. Alundum thimbles may be used and cleaned using standard procedures for glassware. However, disposable cellulose thimbles were used to avoid contamination and the expense of replacing the fragile (and expensive) alundum thimbles.

4.2 *Glass bottles* — Twenty-milliliter glass with foil liners on lids.

4.3 *Balance* — Top loader, capable of weighting 400 g to the nearest 0.001 g.

4.4 *Short-stem funnel.*

4.5 *Magnetic stirrer,* with Teflon-coated stirring bars.

4.6 *Infrared spectrophotometer,* scanning or fixed wavelength, for measurement around 2924 cm (Buck Scientific Model No. HC-404 or equivalent).

4.7 *Cells,* 10-, 50-, and 10-mm path length; quartz; sodium chloride or infrared-grade glass. The 10 mm-cell is typically used.

5 REAGENTS

5.1 *Reagent water* — Reagent water is defined as deionized/distilled water in which an interferant is not observed at the practical quantitation limit of the parameters of interest.

5.2 *Solvents* — Fluorocarbon-113 (Freon), spectrophotometric grade, boiling point 48°C. Methylene chloride, spectrophotometric grade.

5.3 *Sodium sulfate* — Powdered, anhydrous. Store in capped bottle.

5.4 *Silica gel,* 60 to 200 mesh, Davisil Grade 62 Special or equivalent. Should contain 1 to 2% water as defined by residue test at 130°C. Adjust by overnight equilibration if needed.

5.5 *Calibration mixtures*

    5.5.1 *Reference oil* — Pipet 15.0 ml *n*-hexadecane, 15.0 ml isooctane, and 10.0 ml chlorobenzene into a 50-ml glass-stoppered bottle. (A certified premixed reference oil may also be purchased in 1-ml vials [Buck Scientific 404-11]).

    5.5.2 *Stock standard* — Pipet 10 μl reference oil into a tared 20-ml glass vial. Add approximately 10 ml Freon, replace cap, and weigh. The specific gravity of the stock reference solution is 0.831 g/ml and the specific gravity is 1.5635 g/ml. Therefore, this stock standard will have an approximate concentration of 830 mg/l. *Note:* due the volatility of Freon, caps should be sealed at all times except when transfers are being made. Also, the volatility makes accurate volumetric transfer nearly impossible. Therefore, we deal in terms of mass.

    5.5.3 *Working analytical standards* — Transfer approximately 10 ml Freon to a tared glass vial. Replace cap. Transfer aliquot (from 0.25 to 2 ml) of stock oil reference. Cap and weigh. It is desirable that the absorbance of the standards fall between 0.1 and 0.8 absorbance units.

    5.5.4 *Spike mix standard* — Spike standards should be prepared from a source independent of the working standards. Pipet 1 ml of approximately 1000-mg/l stock solution into 20-ml vial. Add 5 ml Freon. The sample should be spiked at a minimum of 50% of the sample matrix concentration. The amount of spike may need to be adjusted on an individual sample basis.

    Note: All preparations should be checked for integrity before usage to ensure appropriate concentrations.

5.6 *Pyrex wool*

6 SAMPLE COLLECTION, PRESERVATION, AND HANDLING

6.1   *Introduction* — Once the sample has been collected, it must be stored and preserved to maintain the chemical and physical properties that it possessed at the time of collection. The sample type, types of containers and their preparation, possible forms of contamination, and preservation methods are all items which must be thoroughly examined in order to maintain the integrity of the samples.

6.2   *Sample handling and preservation* — Types of sample containers and sample preservation should be considered at time of sample collection and should have already been designated before the samples arrive at the laboratory. All samples should be assumed to be hazardous and should be treated as such.

   For soil samples, a representative sample should be collected in a glass container. If analysis is to be delayed more than a few hours, the sample should be refrigerated at 4°C. All samples and extracts should be stored at 4°C to maintain their integrity.

6.2.1   *Drying and grinding procedure for soil samples: rationale* — Experimental observations in the Kansas State University Soil Chemistry Laboratory clearly demonstrate the following problems with standard protocols: (a) It is very difficult to obtain reproducible subsamples from field-moist soils. This is because moist soils do not homogenize easily. (b) Soxhlet extraction of soils generates variability in total petroleum hydrocarbon (TPH) and target compounds. The lack of intimate contact between the solvent and the interior surfaces of the soil is the source of this problem. (c) Extraction of moist soils with nonpolar solvents can be inefficient. If the extraction does not interact with the polar surfaces of the soil or is hydrophobic, then efficiencies as low as 10% are likely with all but sandy soils. To overcome these problems, we are developing new protocols for sample preparation and extraction. These protocols will be thoroughly tested and peer reviewed before implementation. Standard procedures will be used in parallel.

6.2.2   *Drying procedure for soil samples* — All soils will be air dried and ground to pass a 40-mesh sieve. Air-dried, ground soils can be homogenized and yield little subsampling error (<1%) even on 0.5-g samples. A potential limitation to this procedure is the loss of volatiles. Nearly all the volatiles will be lost from the field site with its shallow soils and frequent tillage, but some will be present early in the project. This problem will be quantified as part of the study.

1.   Homogenize the field-moist sample as much as possible.
2.   Transfer approximately 25% of the sample to a clean bottle that has been marked with the appropriate laboratory number. Archive moist sample in refrigerator at 4°C.
3.   Transfer the remaining 75% of the moist sample to a labeled pan and place in forced-air drying oven at ambient temperature.
4.   After 48 hr or when weight is constant, transfer sample back onto appropriately labeled sample bag for grinding.

6.2.3   *Grinding procedure for soil samples*

1.   Transfer dry soil into ceramic mortar.
2.   Grind with mortar and pestle.
3.   Transfer to 40-mesh screen (brass) with brass catch pan.
4.   Place those particles that do not pass the 40-mesh screen back to mortar and grind until all particles pass sieve.
5.   Transfer sample back to sample bag.

6.3   *Safety*
   - Standard laboratory safety precautions should be adhered to at all times. This assumes that all samples are hazardous. The use of hoods, safety glasses, aprons, gloves, and lab coats is mandatory.
   - All operations involving solvents must be performed in a hood.
   - Do not use cracked glassware. Dispose of or have chipped and broken glassware repaired.
   - Excess solvents are emptied into a solvent waste can stored under a hood. When possible, clean and recycle the solvents.
   - Knowledge and use of the Safety Manual is mandatory.

6.4   *Cleaning of glassware* — The basic cleaning steps are
   1. Rinse the glassware with water soon after usage. Follow this with a solvent wash, methylene chloride:acetone (1:1, v/v).
   2. Wash glassware in commercial dishwasher or in sink with hot, soapy water.
   3. After washing, rinse with hot tap water.
   4. Rinse glassware with distilled, deionized water. Follow this with a methanol rinse and a methylene chloride rinse.
   5. Air-dry glassware.
   6. Place glassware in the kiln and bake at 950°C for at least 1 hour.
   7. Allow glassware to cool and cover with aluminum foil before storing.

7 EXTRACTION PROCEDURE

The first procedure described is the extraction of soil by standard soxhlet, SW-846 Method 3540A. The silica gel cleanup is SW-846 Method 3630A.

7.1   Record sample numbers to be extracted in the TPH Laboratory logbook along with blanks and quality control samples.

7.2   Mix sample thoroughly, especially composite samples. Discard any foreign objects, such as leaves, sticks, and rock.

7.3   Label a clean 500-ml soxhlet flask with the job and sample number.

7.4   Place the flask on the balance. Use cork ring or equivalent to support flask. Add a subsample of approximately 5 g to flask. Record mass to nearest 0.001 g. Mix the sample with approximately 5 g of anhydrous sodium sulfate (5.3) until the sample has a sandy, free-flowing texture, adding more if necessary. A blank will consist only of 5 g sodium sulfate.

7.5   If the sample is a matrix spike, matrix spike duplicate, blank spike, or blank spike duplicate, use a disposable 1.0-ml serological pipette to add 1.0 ml of the spike mix (5.6.4) to the sample. The sample should be spiked at a minimum of 50% of the same matrix concentration. The amount of spike may need to be adjusted on an individual sample basis.

7.6   Immediately add 100 ml of methylene chloride to the flask.

7.7   Connect the flask to the soxhlet apparatus and attach the heating mantle. Begin the flow of water through the condenser. Turn on power to the mantle at a low setting. The process must go through at least 10 reflux/siphon cycles (our results suggest that all the TPH is extracted after six cycles).

7.8   Turn off the power and allow the lower flask to cool. Transfer the extract (including the remaining liquid surrounding the thimble) to a clean and labeled 20-ml vial. Seal with a foil-lined cap.

7.9   Transfer 1.0 ml of the methylene chloride to a glass vial in a hood. Allow the methylene chloride to evaporate completely. Evaporation usually takes less than 5 min, but allow 10 min because any remaining methylene chloride will have a strong positive interference in the infrared determination.

7.10   Add approximately 5 ml Freon to the vial. Cap and resuspend the TPHs in the Freon. Add 1 g silica gel, and shake to remove polar organic compounds.

7.11   Filter the suspension through a rapid paper (such as Whatman 41) into a tared 20-ml vial. Cap and reweigh.

7.12   Fill an infrared cuvette with the extract and read the absorbance.

Notes:

1. The size of the aliquot will be dependent upon the concentration of TPH in the solution. Ideally, the absorbance reading will be between 0.1 and 0.8 absorbance units. Smaller or larger aliquots may be needed, or dilutions may be required.

2. The silica gel serves to remove the polar hydrocarbons, such as vegetable oils and animal fats, in order to analyze for petroleum hydrocarbons (mineral oils). The amount of silica gel used may vary and should be checked for each matrix.

7.13   Select the appropriate working standards and cell path length according to the desired analysis range. Calibrate the instrument for the appropriate cells using a series of five working standards (5.6.3). It is not necessary to add silica gel to the standards. Determine the absorbance directly for each solution at the absorbance maximum between 2940 and 2900 cm^{-1}. Prepare the calibration plot of absorbance vs. milligrams petroleum hydrocarbons per 100 ml solution.

7.14   Fill a clean cell with sample solution and determine the absorbance of the extract. If the absorbance exceeds that of the most concentrated standard, prepare the appropriate dilution.

Note: The possibility that the absorptive capacity of the silica gel has been exceeded can be tested at this point by adding another 3.0 g silica gel to the extract and repeating the treatment and determination.

8 CALCULATIONS

8.1   Calculate the petroleum hydrocarbons in the sample using the formula

$$R * D/W$$

where R = milligrams of petroleum hydrocarbons as determined from the calibration plot (7.14); D = extract dilution factor, if used; and W = weight of sample, in kilograms.

9 QUALITY CONTROL

9.1   All extractions of soil samples should be within 14 days of collection.

9.2   The calibration of the infrared spectrophotometer should be checked daily.

9.3   All spiking solutions should be checked for integrity before usage.

9.4   All solvent lots should be checked for purity and usefulness by running a total blank on the lot.

9.5   Extract a blank for every set of 20 samples with a minimum of one blank per set of samples.

9.6   A matrix spike and matrix spike duplicate will be extracted every 20 samples. In addition, a method blank spike (LCS) and method blank spike duplicate (LCSD) should be extracted for every 20 samples. If there is insufficient sample or all samples are too concentrated for spike recoveries to be observed, the LCS and LCSD recovery values should be reported.

10 RELATIVE PERCENT RECOVERY (RPR) AND RELATIVE PERCENT DIFFERENCE (RPD) LIMITS

Laboratory control samples must meet the criteria below. Sample spikes and replicates which do not meet these limits should be reextracted once, adjusting the spike concentration if necessary. If the sample RPRs and/or RPDs still do not pass, a corrective action report should be included with the report. SW846 indicates that samples that exceed 0.1% (1000 ppm) need not be spiked. In this case, the LCS data should be reported with a comment in the report to that effect.

LCS RPR	73–119%
LCS RPD	0–14%
Sample matrix RPR	45–155%
Sample matrix RPD	0–26%

11 TPH RUN LOGBOOK

The following information must be included for all entries in the TPH laboratory logbook:

Date	Cell length
Analyst	Analysis
File #	Matrix
Work order	Dilution
Dash	Comments
Sample ID	Batch #
Scan range	

REFERENCES

Petroleum Hydrocarbons, Total Recoverable, EPA 418.1 (Spectrophotometric, infrared), STORET No. 45501.

Research and Engineering Safety Manual. 1992. Department of Public Safety, Edwards Hall, Kansas State University.

Standard Extraction for Base, Neutral and Acid Pollutants from Soil by SW-846, Method 3550 Sonication Extraction, EVSD-EPA3550-002-90.

U.S. Environmental Protection Agency. 1984. Test Methods for Evaluating Solid Waste — Physical Chemical Methods, 3rd ed., Publ. SW-846, Office of Solid Waste, Washington, D.C.

Full-Scale Design Drawings and Calculations

Table A2.1 Basis of Design

Parameter	Value	Used in Calculating
Area	13.77 acres (5.58 ha)	Total costs, size of irrigation system, and duration of construction
Depth of soil	2 ft (0.6 m)	Total costs and duration of construction
Grade	0 to 5%	Size of irrigation system
Infiltration rate	0.5 in./hr (1.3 cm/hr)	Size of irrigation system
Application rate	3 in. (7.6 cm) per application	Size of irrigation system
Average wind speed	7 to 10 mi/hr (11 to 16 km/hr)	Size of irrigation system

Table A2.2 Full-Scale Design: Advanced Applied Technology Development Facility — Craney Island

1.	Area	Dimensions same as Craney Island biotreatment facility 600 ft × 1000 ft = 600,000 ft²/43,560 　　　　　　= 13.77 acres
2.	Depth	Use 2-ft depth as in study
3.	Volume	600 ft × 1000 ft × 2 ft/27 ft³/yd³ ≅ 45,000 yd³
4.	Irrigation system design	Use solid set system for low operation and maintenance costs
4.1	Application rate, I	I < infiltration rate of bare soil, to prevent runoff and uneven distribution
	Design conditions 　Grade	Assume grade 0 to 5% No application rate reduction
	Soil	Loam, infiltration = 0.5 in./hr (assumed value)
	Depth of water	3 in. per application
	Application area	13.77 acres
	Average wind speed	7 to 10 mi/hr
4.2	Design constraints	Avoid pipes at entrance Minimize number of laterals Use readily available agricultural sprinklers
4.3	Select lateral and sprinkler spacings	S_L = 150 ft S_S = 200 ft

145

Table A2.2 Full-Scale Design: AATDF — Craney Island (continued)

4.4	Calculate required sprinkler discharge, equation E-7 (EPA, 1981)	$QS = \dfrac{IS_S S_L}{C}$ = <0.5 in./hr 200 ft 150 ft/96.3 $QS = <156 \cong 150$ gpm
4.5	Select sprinkler head using manufacturer's data (see Rainbird data, attached)	"Raingun" 2-in. inlet 1.0-in. nozzle GNS-2T Wetted diameter = 216 ft, 152 gpm @ 30 psi Provide overlapping coverage for wind condition
4.6	System flow capacity, Q	$Q = A\,I = 13.77$ acres × 0.5 in./hr $\dfrac{27.156\ \text{gal}}{\text{acre-inch}} \times \dfrac{1\ \text{hr}}{60\ \text{min}}$ = 3116 gal/min
4.7	Application periods	$T = D/I$ $I = 0.5$ in./hr

D (in.)	T (hr)
1	2
2	4
3	6
4	8

4.8	Irrigation piping size	Pipe sizes estimated using Jack's cube equations (*Chem. Eng. Prog.*, December 1997) $Q = 1.2\,(D + 2)^3$ Use standard PVC, schedule 40 $P = 30$ psi Layout shown in Figure A2.2
	Main header Laterals Risers	$Q = 3116$ gal/min, $D = 12$ in. $Q = 3116/5$ laterals = 623 gal/min, $D = 6$ in. Height = 3 ft, $D = 2$ in., per Rainbird specs

Provide manual gate valves at main and at each lateral to control flow

5. Site work
  5.1 Load and haul contaminated soil from stockpiles and spread 2 ft over treatment area

  Equipment: load using wheel loaders, 5-yd³ capacity

  Haul and spread using bottom-dump scaper; it spreads 10 ft wide, 0.7 ft thick, 100 ft long; 34-yd³ capacity (Church, 1981)[a]

  Number of runs = 44,444 yd³/34 yd³/run = 1308

  Total hours = 1308 × 0.5 hr per run to nearby stockpile
  = 654 hr

  5.2 Install irrigation system
  Duration = two calendar weeks (Atlantic Irrigation, 1997)

  5.3 Apply lime

  5.4 Hydroseed with initial fertilizer/mulch mix

[a] Church, H.K. 1981. *Excavation Handbook*, McGraw-Hill, New York.

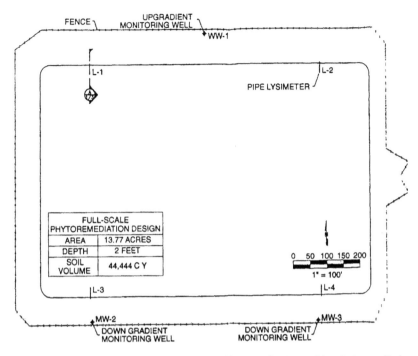

Figure A2.1 Full-scale phytoremediation plan view with groundwater and leachate monitoring.

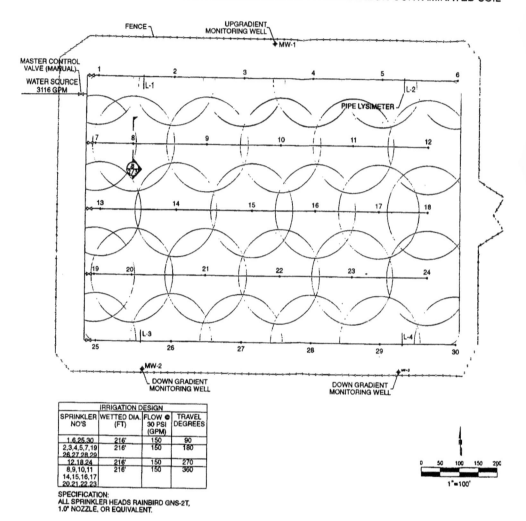

IRRIGATION DESIGN			
SPRINKLER NO'S	WETTED DIA. (FT)	FLOW @ 30 PSI (GPM)	TRAVEL DEGREES
1,6,25,30	216'	150	90
2,3,4,5,7,19 26,27,28,29	216'	150	180
12,18,24	216'	150	270
8,9,10,11 14,15,16,17 20,21,22,23	216'	150	360

SPECIFICATION:
ALL SPRINKLER HEADS RAINBIRD GNS-2T,
1.0" NOZZLE, OR EQUIVALENT.

Figure A2.2 Full-scale phytoremediation irrigation plan.

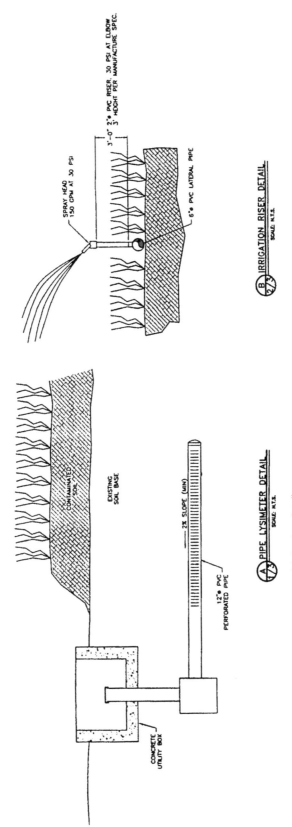

Figure A2.3 Full-scale remediation lysimeter and irrigation details.

Project Cost Spreadsheets

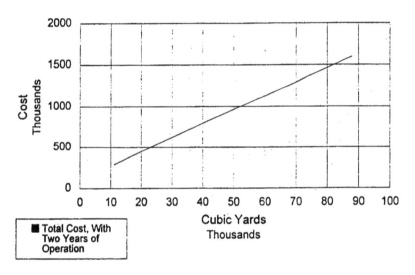

Figure A3.1 Total cost vs. soil volume.

Summary

Years of Operation	Acres	Total Cost ($)					
		1	2	3	4	5	6
Phytoremediation scale 1	3	252,812	296,635	340,458	384,281	428,103	471,926
Phytoremediation scale 2	7	429,029	494,204	559,380	624,555	689,731	754,906
Phytoremediation scale 3	14	782,010	892,479	1,002,949	1,113,418	1,223,887	1,334,356
Phytoremediation scale 4	27	1,412,631	1,605,923	1,799,215	1,992,507	2,185,799	2,379,091

Remediation Technologies, Inc.
Biotreatment Alternatives Review

	Phytoremediation Scale 1	16,500 tons 11,000 yd^3	24-in. depth 3.4 acres		
Tasks	Description	Units	Unit Cost ($)	Total Cost ($)	Note
100	Project contracts, permits, and reports (except operations)				
	Personnel				
	Project manager	120 hr	95.00	11,400	1
	Site engineer	40 hr	70.00	2,800	1
	Cost analysis	5 hr	55.00	275	1
	Secretary	20 hr	40.00	800	1
	Travel				
	Per diem, lodging, rental car	10 days	150.00	1,500	1
	Materials				
	Phone, fax, photocopy, postage	1 lump sum	500.00	500	1
	Task subtotal			$17,275	
200	Project engineering (total)				
	Labor	10% of capital		16,474	1
	Materials	1 lump sum	500.00	500	1
	Plant selection study	1 lump sum	10,000.00	10,000	1
	Task subtotal			$26,974	
300	Construction (total)				
	Mobilization				
	Facilities transport and hookup	1 lump sum	800.00	800	1
	Equipment transport	1 lump sum	2,000.00	2,000	1
	Site prep	2 days	800.00	1,600	1
	Treatment area				
	Ramp/berm construction	1,541 linear feet	5.60	8,632	2
	Irrigation system	1.00 lump sum	30,378.00	30,378	3, 7
	Soil placement				
	Load, haul, spread	11,000 yd^3	10.00	110,000	2
	Initial tilling				
	with lime addition @ 1 acre/hr	3.41 acres	227.00	774	2
	Demobilization				
	Facilities transport	1 lump sum	450.00	450	1
	Equipment transport	1 lump sum	1,800.00	1,800	1
	Site cleanup	2 days	1,000.00	2,000	1
	Task subtotal			$158,434	
400	Phytoremediation hydroseeding				
	Hydroseed at start of project	3.41 acres	1,850.00	6,307	2
	Task subtotal			$6,307	
Total	Total capital cost			$208,990	
500	Operations (per year)				
	Equipment				
	Tractor/harvester (2× per season)	3.41 acres	2,000.00	6,818	4
	Labor				
	Site engineer	80 hr	88.00	7,040	1
	Field technicians/equipment operators	100 hr	40.00	4,000	1
	Health and safety				
	Protective equipment/monitoring	24 days	150.00	3,600	1
	Field office				
	Office trailer/facilities	12 months	302.50	3,630	1
	Electricity	12 months	50.00	600	1
	Daily expenses				
	Per diem, lodging, transport, shipping	24 days	150.00	3,600	1
	Fertilizer				
	Rotary application	3.41 acres	82.00	280	2
	Water				
	Irrigation (3 in./week × 10 weeks)	2,777 Thsnd Gal	0.100	278	5
	Task subtotal			$29,845	

Tasks	Description	Units	Unit Cost ($)	Total Cost ($)	Note
600	**Analytical program**				
	Off-site analyses				
	(per year)	3.41 acres	4,100.00	13,977	6
	Task subtotal			$13,977	
Total	**Total annual operations cost**			$43,823	
Liner	**Optional prepared bed**				
	Prep treatment bed: scrape top 4 in.	0 yd^3	5.03	0	2
	Liner	1 lump sum	259,919.00	259,919	2, 7
	Leachate collection system	8 linear feet	4.85	39	2
	Additional engineering	10% of capital		25,996	
	Total			$285,954	

Remediation Technologies, Inc.
Biotreatment Alternatives Review

	Phytoremediation Scale 2	33,000 tons 22,000 yd3	24-in. depth 6.8 acres		

Tasks	Description	Units	Unit Cost ($)	Total Cost ($)	Note
100	**Project contracts, permits, and reports (except operations)**				
	Personnel				
	Project manager	120 hr	95.00	11,400	1
	Site engineer	40 hr	70.00	2,800	1
	Cost analysis	5 hr	55.00	275	1
	Secretary	20 hr	40.00	800	1
	Travel				
	Per diem, lodging, rental car	10 days	150.00	1,500	1
	Materials				
	Phone, fax, photocopy, postage	1 lump sum	500.00	500	1
	Task subtotal			$17,275	
200	**Project engineering (total)**				
	Labor	10% of capital		30,553	1
	Materials	1 lump sum	500.00	500	1
	Plant selection study	1 lump sum	10,000.00	10,000	1
	Task subtotal			$41,053	
300	**Construction (total)**				
	Mobilization				
	Facilities transport and hookup	1 lump sum	800.00	800	1
	Equipment transport	1 lump sum	2,000.00	2,000	1
	Site prep	2 days	800.00	1,600	1
	Treatment area				
	Ramp/berm construction	2,180 linear feet	5.60	12,207	2
	Irrigation system	1.00 lump sum	50,507.00	50,507	3, 7
	Soil placement				
	Load, haul, spread	22,000 yd^3	10.00	220,000	2
	Initial tilling	6.82 acres	227.00	1,548	2
	with lime addition @ 1 acre/hr				
	Demobilization				
	Facilities transport	1 lump sum	450.00	450	1
	Equipment transport	1 lump sum	1,800.00	1,800	1
	Site cleanup	2 days	1,000.00	2,000	1
	Task subtotal			$292,912	
400	**Phytoremediation hydroseeding**				
	Hydroseed at start of project	6.82 acres	1,850.00	12,614	2
	Task subtotal			$12,614	
Total	**Total capital cost**			$363,853	

Tasks	Description	Units	Unit Cost ($)	Total Cost ($)	Note
500	**Operations (per year)**				
	Equipment				
	Tractor/harvestor (2× per season)	6.82 acres	2,000.00	13,636	4
	Labor				
	Site engineer	80 hr	88.00	7,040	1
	Field technicians/equipment operators	100 hr	40.00	4,000	1
	Health and safety				
	Protective equipment/monitoring	24 days	150.00	3,600	1
	Field office				
	Office trailer/facilities	12 months	302.50	3,630	1
	Electricity	12 months	50.00	600	1
	Daily expenses				
	Per diem, lodging, transport, shipping	24 days	150.00	3,600	1
	Fertilizer				
	Rotary application	6.82 acres	82.00	559	2
	Water				
	Irrigation (3 in./week × 10 weeks)	5,555 Thsnd Gal	0.100	555	5
	Task subtotal			$37,221	
600	**Analytical program**				
	Off-site analyses				
	(per year)	6.82 acres	4,100.00	27,955	6
	Task subtotal			$27,955	
Total	**Total annual operations cost**			$65,175	
Liner	**Optional prepared bed**				
	Prep treatment bed: scrape top 4 in.	0 yd³	5.03	0	2
	Liner	1 lump sum	432,139.00	432,139	2, 7
	Leachate collection system	8 linear feet	4.85	39	2
	Additional engineering	10% of capital		43,218	
	Total			$475,396	

Remediation Technologies, Inc.
Biotreatment Alternatives Review

	Phytoremediation Scale 3	68,000 tons 45,333 yd³	24-in. depth 14.0 acres		
Tasks	**Description**	**Units**	**Unit Cost ($)**	**Total Cost ($)**	**Note**
100	**Project contracts, permits, and reports (except operations)**				
	Personnel				
	Project manager	120 hr	95.00	11,400	1
	Site engineer	40 hr	70.00	2,800	1
	Cost analysis	5 hr	55.00	275	1
	Secretary	20 hr	40.00	800	1
	Travel				
	Per diem, lodging, rental car	10 days	150.00	1,500	1
	Materials				
	Phone, fax, photocopy, postage	1 lump sum	500.00	500	1
	Task subtotal			$17,275	
200	**Project engineering (total)**				
	Labor	10% of capital		58,524	1
	Materials	1 lump sum	500.00	500	1
	Plant selection study	1 lump sum	10,000.00	10,000	1
	Task subtotal			$69,024	
300	**Construction (total)**				
	Mobilization				
	Facilities transport and hookup	1 lump sum	800.00	800	1
	Equipment transport	1 lump sum	2,000.00	2,000	1
	Site prep	2 days	800.00	1,600	1

Tasks	Description	Units	Unit Cost ($)	Total Cost ($)	Note
	Treatment area				
	Ramp/berm construction	3,129 linear feet	5.60	17,524	2
	Irrigation system	1.00 lump sum	76,554.00	76,554	3
	Soil placement				
	Load, haul, spread	45,333 yd^3	10.00	453,333	2
	Initial tilling	14.05 acres	227.00	3,189	2
	with lime addition @ 1 acre/hr				
	Demobilization				
	Facilities transport	1 lump sum	450.00	450	1
	Equipment transport	1 lump sum	1,800.00	1,800	1
	Site cleanup	2 days	1,000.00	2,000	1
	Task subtotal			$559,250	
400	**Phytoremediation hydroseeding**				
	Hydroseed at start of project	14.05 acres	1,850.00	25,992	2
	Task subtotal			$25,992	
Total	**Total capital cost**			$671,541	
500	**Operations (per year)**				
	Equipment				
	Tractor/harvestor (2× per season)	14.05 acres	2,000.00	28,099	4
	Labor				
	Site engineer	80 hr	88.00	7,040	1
	Field technicians/equipment operators	100 hr	40.00	4,000	1
	Health and safety				
	Protective equipment/monitoring	24 days	150.00	3,600	1
	Field office				
	Office trailer/facilities	12 months	302.50	3,630	1
	Electricity	12 months	50.00	600	1
	Daily expenses				
	Per diem, lodging, transport, shipping	24 days	150.00	3,600	1
	Fertilizer				
	Rotary application	14.05 acres	82.00	1,152	2
	Water				
	Irrigation (3 in./week × 10 weeks)	11,446 Thsnd Gal	0.100	1,145	5
	Task subtotal			$52,866	
600	**Analytical program**				
	Off-site analyses				
	(per year)	14.05 acres	4,100.00	57,603	6
	Task subtotal			$57,603	
Total	**Total annual operations cost**			$110,469	
Liner	**Optional prepared bed**				
	Prep treatment bed: scrape top 4 in.	7,556 yd^3	5.03	38,004	2
	Liner	612,000 ft^2	1.07	654,840	2
	Leachate collection system	6,258 linear feet	4.85	30,353	2
	Additional engineering	10% of capital		72,320	
	Total			$795,518	

Remediation Technologies, Inc.
Biotreatment Alternatives Review

	Phytoremediation Scale 4	132,000 tons 88,000 cubic yards	24-in. depth 27.3 acres		
Tasks	Description	Units	Unit Cost ($)	Total Cost ($)	Note
100	**Project contracts, permits & reports (except operations)**				
	Personnel				
	Project manager	120 hr	95.00	11,400	1
	Site engineer	40 hr	70.00	2,800	1

Tasks	Description	Units	Unit Cost ($)	Total Cost ($)	Note
	Cost analysis	5 hr	55.00	275	1
	Secretary	20 hr	40.00	800	1
	Travel				
	Per diem, lodging, rental car	10 days	150.00	1,500	1
	Materials				
	Phone, fax, photocopy, postage	1 lump sum	500.00	500	1
	Task subtotal			$17,275	
200	**Project engineering (total)**				
	Labor	10% of capital		108,324	1
	Materials	1 lump sum	500.00	500	1
	Plant selection study	1 lump sum	10,000.00	10,000	1
	Task subtotal			$118,824	
300	**Construction (total)**				
	Mobilization				
	Facilities transport and hookup	1 lump sum	800.00	800	1
	Equipment transport	1 lump sum	2,000.00	2,000	1
	Site prep	2 days	800.00	1,600	1
	Treatment area				
	Ramp/berm construction	4,360 linear feet	5.60	24,415	2
	Irrigation system	1.00 lump sum	113,530.00	113,530	3, 7
	Soil placement				
	Load, haul, spread	88,000 yd^3	10.00	880,000	2
	Initial tilling	27.27 acres	227.00	6,191	2
	with lime addition @ 1 acre/hr				
	Demobilization				
	Facilities transport	1 lump sum	450.00	450	1
	Equipment transport	1 lump sum	1,800.00	1,800	1
	Site cleanup	2 days	1,000.00	2,000	1
	Task subtotal			$1,032,786	
400	**Phytoremediation hydroseeding**				
	Hydroseed at start of project	27.27 acres	1,850.00	50,455	2
	Task subtotal			$50,455	
Total	**Total capital cost**			$1,219,339	
500	**Operations (per year)**				
	Equipment				
	Tractor/harvestor (2× per season)	27.27 acres	2,000.00	54,545	4
	Labor				
	Site engineer	80 hr	88.00	7,040	1
	Field technicians/equipment operators	100 hr	40.00	4,000	1
	Health and safety				
	Protective equipment/monitoring	24 days	150.00	3,600	1
	Field office				
	Office trailer/facilities	12 months	302.50	3,630	1
	Electricity	12 months	50.00	600	1
	Daily expenses				
	Per diem, lodging, transport, shipping	24 days	150.00	3,600	1
	Fertilizer				
	Rotary application	27.27 acres	82.00	2,236	2
	Water				
	Irrigation (3 in./week × 10 weeks)	22,219 Thsnd Gal	0.100	2,222	5
	Task subtotal			$81,474	
600	**Analytical program**				
	Off-site analyses				
	(per year)	27.27 acres	4,100.00	111,818	6
	Task subtotal			$111,818	
Total	**Total annual operations cost**			$193,292	
Liner	**Optional prepared bed**				
	Prep treatment bed: scrape top 4 in.	0 yd^3	5.03	0	2

Tasks	Description	Units	Unit Cost ($)	Total Cost ($)	Note
	Liner	1 lump sum	971,366.00	971,366	2, 7
	Leachate collection system	8 linear feet	4.85	39	2
	Additional engineering	10% of capital		97,140	
	Total			$1,068,545	

[1] Costs for professional labor associated with permitting, engineering, sampling, and management were estimated from general RETEC project experience.

[2] Costs for site work obtained from two R.S. Means Construction Publishers & Consultants (Kingston, MA) Cost Guides: Environmental Restoration Assemblies Cost Book, 1997 and Means Site Work & Landscape Cost Data, 1995. These costs include materials, installation labor, overhead, and profit and assume work in Level D conditions by workers with OSHA HazWoper training.

[3] The irrigation system unit costs were derived from the cost of the irrigation system designed for the Craney Island full-scale system (13.8 acres), as follows:

Craney Island Phytoremediation Full-Scale Design
Irrigation System Costs

Item	Spec	Quantity	Units	Unit Price ($)	Total Price ($)	Reference
Header	12-in. PVC, Sch. 40	600	ft	8.5	5,100	Rochester Irrigation and
Lateral	6-in. PVC, Sch. 40	4,700	ft	2.4	11,280	Plumbing Supply
Riser	2-in. PVC, Sch. 40	90	ft	0.6	54	(Mr. Robert Tyndall 716-232-6770)
Fittings						
Elbow	12/6-in. PVC, Sch. 40	6	ea	278	1,668	"
Tee	12-in. PVC, Sch. 40	1	ea	382	382	"
Tee	12/6-in. PVC, Sch. 40	3	ea	275	825	"
Tee	6/2-in. PVC, Sch. 40	30	ea	30	900	"
Valve	B-fly 12-in. PVC, Sch. 40	1	ea	967	967	"
Valve	B-fly 6-in. PVC, Sch. 40	5	ea	384	1,920	"
Sprinkler	Rainbird GNS-2T 1.0-in. nozzle					Rainbird Eastern Representative
	Full circle F2002	15	ea	670	10,050	Mr. Claude Corcoris
	Partial SR 2003	15	ea	980	14,700	(732-748-0222)
Installation	Factor of 0.5 × materials	1	lump sum		23,923	Rochester Irrigation and Plumbing Supply
Start-up/ controls	System (see Figure A2.2)	1	lump sum		4,785	10% of capital
TOTAL					76,554	
Unit cost		1	acre		5,629	Round $5,600

[4] Tractor/harvester costs are from local (Virginia) contractor (includes mobilization/demobilization).

[5] Cost of water derived from local water municipal supply costs of $0.10 per 1000 gal.

[6] Cost of hydrocarbon analyses from Lancaster Laboratories, PA:

Total petroleum hydrocarbon	2 rounds per season	3 samples per acres	$125	$750
Polycyclic aromatic hydrocarbon	2 rounds per season	3 samples per acre	$225	$1,350
Total				$2,100
Cost of bioavailability and toxicity testing from Gas Research Institute's EAE program				$2,000
Total				$4,100

[7] Cost scaling: A cost scaling technique was used to estimate the costs of irrigation equipment and the cost of the optional treatment bed liner. The six-tenths factor rule was used (Jelen's Cost and Optimization Engineering, Humphreys, K.H., Ed., 1991, Section 14.10, p. 382). Cost 2 = Cost 1(Capacity 2/Capacity 1) exp. 0.6.

1. Irrigation System. The system was designed and costed at the Craney Island full scale (14 acres).

Cost curve
for other scales:

Area (acres)	Cost
3	30,378
5	41,274
7	50,507
10	62,559
12	69,791
14	76,554
16	82,940
18	89,013
20	94,822
22	100,403
24	105,783
26	110,988
27	113,530
30	120,938
34	130,370
38	139,367
42	147,993
46	156,295

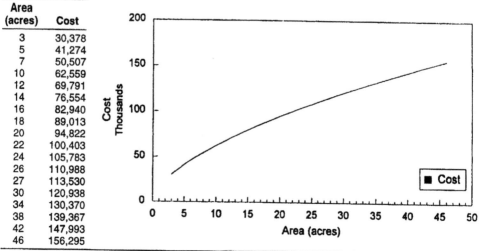

Figure A3.2 Irrigation system cost scaling.

1. Liner System. The system was designed and costed at the Craney Island full scale (14 acres).

Cost curve
for other scales:

Area (acres)	Cost
3	259,919
5	353,140
7	432,139
10	535,260
12	597,136
14	655,000
16	709,637
18	761,602
20	811,302
22	859,049
24	905,089
26	949,617
27	971,366
30	1,034,754
34	1,115,454
38	1,192,435
42	1,266,234
46	1,337,270

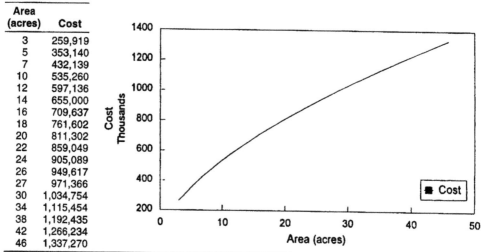

Figure A3.3 Synthetic liner cost scaling.

Index